Cybersecurity Awareness Among Students and Faculty

Cybersecurity Awareness Among Students and Faculty

Abbas Moallem

CRC Press
Taylor & Francis Group
Boca Raton London New York

CRC Press is an imprint of the
Taylor & Francis Group, an **informa** business

CRC Press
Taylor & Francis Group
6000 Broken Sound Parkway NW, Suite 300
Boca Raton, FL 33487-2742

Library of Congress Cataloging-in-Publication Data

Names: Moallem, Abbas, author.
Title: Cybersecurity awareness among students and faculty / by Abbas Moallem.
Description: Boca Raton, FL : CRC Press/Taylor & Francis Group, 2019. |
Includes bibliographical references.
Identifiers: LCCN 2019009156 | ISBN 9780367144074 (hardback : alk. paper) |
ISBN 9780429031908 (ebook)
Subjects: LCSH: Computer security. | Computer networks—Security measures. |
Universities and colleges—Security measures.
Classification: LCC QA76.9.A25 M654 2019 | DDC 005.8—dc23
LC record available at https://lccn.loc.gov/2019009156

Visit the Taylor & Francis Web site at
http://www.taylorandfrancis.com

and the CRC Press Web site at
http://www.crcpress.com

To my wife, Parnian Taidi, and my daughters, Shabnam and Nassim Moallem.

Contents

Preface

The digital world has fundamentally changed the way of life of all human beings, and continuing changes can be expected in the future. Sooner than one might think, all societies, rich and poor, will fully interface with the Internet. From shopping, banking, and government affairs, to meetings, connecting with people, and even playing games, all humankind will soon be connected through network and computer interfaces. The resulting data shared over the network or produced and stored in the global systems of interlinked servers will be saved forever. Data from members, customers, users, or visitors will be accessible to whoever owns it. We can expect that both legal and, unfortunately, illegal sharing of stored data will occur regularly and all too frequently. In addition to transactional and search data, the deployment of surveillance cameras, which are already widely used for security and management purposes, will expand everywhere, using drones and personal surveillance systems. We can clearly foresee extraordinary systems of surveillance of every human being in the very near future.

Soon, every single home will be equipped with multiple Internet-enabled devices capable of interacting with each other over the network. A tremendous number of electric and smart cars or self-driving cars will be on the road. Thus, while the consequences of hacking have so far been limited to the digital world, now hacking will affect the physical world as well. For example, if a hacker penetrates a bank account, a victim will lose his/her money. If a hacker gets access to a smart or self-driving car, a victim may lose his/her life. There are also similar problems with industrial installations, in particular critical infrastructure. Thus, the vulnerability of the new system and management of its security will not be just finding a technical solution but safeguarding the physical, digital property, and privacy of the system.

Along with this development, hacking incursions of database storage systems perpetrated by a variety of agents and agencies (criminals, spying agencies, litigation, etc.) for a wide range of reasons (financial gain, political influence, information, intellectual property trade secret, revenge, hate, desire to inflict harm, etc.) will continue to grow. In fact, it is unsettling but expected that cyber-attacks will grow much faster than finding technological solutions to fully protect individuals, organizations, and societies.

Consequently, in modern times, all individuals must not only be aware and knowledgeable about cybersecurity but also have practical skills and abilities to protect themselves in cyberspace. People have consistent behaviours in locking their homes or putting their valuables in the safe deposit box of a bank or at home. They now must learn to do the same for their digital assets.

It is essential for a proper action plan to begin with measuring cybersecurity knowledge and getting people to extensively protect themselves in cyberspace if they currently have only a limited or fair knowledge of cybersecurity. However, measuring cybersecurity knowledge at a global level is complex, due to the diversity of communities in age, level of education, geographical location, standard of living, and technology usage. Consequently, the approach taken in this research was to narrow these variations by evaluating cybersecurity awareness among college students. College students represent the most technologically active portion of the population in any society.

I have concentrated on students in public universities in the San Francisco Bay Area of California. Public universities have very diverse student populations comprising different ethnic groups and all income levels. The Bay Area is one of the world's most advanced communities from the perspectives of wealth, technology, progress, and diversity of a population. The aim is that the results of this investigation might help over 207 million college students and 26,000 higher education institutions worldwide to raise their awareness by creating security policies and educational programmes in cybersecurity.

After investigating student awareness in an initial study, I thought it would also be useful to replicate the same study, but this time with faculty members at public universities to have a comparison point.

In this book, after providing background on previous studies, I will summarize the results of several surveys and investigations, and then suggest solutions and a perspective for the future. The results of the investigations summarized in this book might help others to understand student and faculty needs better. The results can also be extrapolated to other organizations and communities as a reference point.

Abbas Moallem

Acknowledgements

I thank Wojciech Cellary, professor at Poznan University, Poznan, Poland; Louis Freud, professor at San Jose State University, San Jose, California; Mahnaz Moallem, professor at Towson University, Towson, Maryland; and Minoo Moallem, professor at the University of California, Berkeley, Berkeley, California, for dedicating their time to review the manuscript of this book and for their valuable suggestions and comments.

I also thank Arash Bayatmakou for editing the manuscript.

Author

Abbas Moallem is an executive director of user experience at UX Experts, LLC, Cupertino, California and an adjunct professor at San Jose State University, San Jose, California, where he teaches courses in cybersecurity, human–computer interaction (HCI), and human factors. Moallem is the editor of the *Human–Computer Interaction and Cyber Security Handbook* published in 2018 by CRC Press.

Dr. Moallem holds a PhD degree in human factors and ergonomics from the University of Paris 13, Paris, France, and has over 30 years of experience in the fields of human factors, HCI, and usability. He has worked with numerous companies including PeopleSoft, Oracle Corporation, Tumbleweed, Axway, NETGEAR, Sears HC, Polycom, Cisco System, HID Global, Lam Research, and Applied Materials.

Introduction

1

1.1 EXTENT OF CYBERATTACKS

In November 2018, the personal data (including credit card details, passport numbers, and dates of birth) of up to 500 million people were stolen in a "colossal" hack of Marriott International, the parent company of hotel chains including W, Westin, Le Méridien, and Sheraton [1]. Two months earlier, in September 2018, press reports surfaced that British Airways had suffered an enormous data breach affecting almost 400,000 customers and including personal and financial details [2]. A month earlier, in August 2018, T-Mobile was hacked and hackers swiped the data of 2 million subscribers [3]. In March 2018, the *New York Times* reported that Cambridge Analytica, a political data firm, gained access to the private information of more than 50 million Facebook users. The firm offered tools that could identify the personalities of American voters and influence their behaviour [4]. Almost a year earlier in September 2017, Equifax, one of three major credit-reporting agencies in the United States, revealed that highly sensitive personal and financial information for about 143 million American consumers was compromised in a cybersecurity breach that began in late spring that year [5].

Every day, cyber-criminals exploit a variety of threat vectors, including email, network traffic, user behaviour, and application traffic to insert ransomware [6]. For example, cyber-criminals use email wiretapping to create an hypertext markup language (HTML) email that, each time it is read, can send back a copy of the email's contents to the originator. It gives the author of the email an opportunity to see to whom the email was subsequently forwarded and any forwarded messages.

Today, technology facilitates communication to the extent that one can chat with someone else in the next room or as far as another country with ease, via a variety of technologies. This ease of communication also prepares the ground for cyber-stalking. Cyber-stalking is defined as the use of technology, particularly the Internet, to harass someone. Typical characteristics include false accusations, monitoring, threats, identity theft, and data destruction or manipulation. Cyber-stalking also includes exploitation of minors, be it sexual or otherwise. Reyns [7] reports that approximately 4.9% of students had perpetrated cyber-stalking in 2009.

Consequently, cybersecurity, or information technology (IT) security, has become one of the major concerns of organizations, communities, and individuals. Cyberspace has become a new site of crime and illegal behaviour. While a wide range of acts of crime and criminality—including robbery, identity theft, ransom, spying, subterfuge, deception, and black markets—have been parts of the experience of social life, globalization and the expansion of new media technologies have presented us with new changes and challenges. With the expansion of digital media, these activities have taken unique forms requiring specific, and sometimes fundamentally distinct, ways of understanding.

1.2 REVIEW OF THE LITERATURE

The cases of cyber-attacks show the extent to which any individual using the Internet and computers is vulnerable to cyber-attacks, which affect not just businesses or organizations but also individuals.

In the following sections, some of the recent studies on students' cybersecurity awareness and privacy will be briefly reviewed.

1.2.1 Cybersecurity Awareness of College Students

User understanding of privacy and security risks, and how to protect themselves from cyber-attacks is a fundamental need in modern life. After all, from banking and e-commerce to pictures of private information and documents, so much can be compromised. Also, information breaches of companies containing user information can easily subject users to identity theft. What users can do to protect themselves and what actions they should take depend on their awareness and knowledge of the risks. The Federal Trade Commission's Consumer Sentinel Network, which collects data about consumer complaints, including identity theft, found that 18% of people who experienced identity theft in 2014 were between the ages of 20 and 29 [8,9].

In recent years, several studies have been conducted to measure the level of awareness among college students concerning information security issues. For example, Slusky and Partow-Navid [10] surveyed students at the College of Business and Economics at California State University, Los Angeles, Los Angeles, California. The results suggested that the major problem with security awareness is not due to a lack of security knowledge but somewhat in the way that students apply that knowledge in real-world situations. Simply put, according to the results of this study, compliance with information security knowledge is lower than the understanding or awareness of it.

Another study conducted by Al-Janabi and Al-Shourbaji [11] analysed cybersecurity awareness among academic staff, researchers, undergraduate students, and employees in the education sector in the Middle East. The results revealed that the participants did not have the requisite knowledge and understanding of the importance of information security principles and their practical applications in day-to-day work.

Hussein and Zhang [12] designed a survey to study the awareness of privacy among a group of users (92% of between 21 and 35, and 76% either engineers or students) who use social media. Their study included 377 participants who use social media services such as Facebook, Twitter, LinkedIn, and Google. The researchers found that 44% of the respondents showed a lack of knowledge of privacy policy and the mechanisms governing it on the online social networks they used. In addition, 34% were gravely concerned, and 41% were somewhat concerned about their privacy online. A staggering 80% indicated that they were not satisfied enough with the level of privacy provided by online social networks.

In a study, Senthilkumar et al. [13] aimed to analyse cybersecurity awareness among college students in Tamil Nadu (a state in India) about various security threats. Five hundred students in five major cities took the online survey. The result showed that 70% of these students were more conscious of basic virus attacks and using antivirus software (updating frequently) or Linux platforms to safeguard their system from virus attacks. The remaining students were not using any antivirus and were the victims of virus attacks. It was also reported that 11% of them were using antivirus but not updating their antivirus software. More than 97% of them did not know the source of the virus.

A study by Grainne et al. [14] was conducted among Malaysian undergraduate students in which 295 took part. The objective was to understand the awareness of risks related to social networking sites (SNSs). The study reported that more than one-third of participants had fallen victim to SNS scams.

1.2.2 Privacy and Self-Disclosure

The Internet and a multitude of social networking applications have massively increased the possibility of the disclosure of personal information. Despite users' concerns and awareness about privacy, their behaviours do not mirror those concerns [15].

Chen et al. [16] discussed a new type of privacy concern, called *Information Privacy Control about Peer Disclosure* (IPCPD). They studied the decisional control to alleviate such a privacy concern by taking certain factors into consideration. "Decisional control" is defined as the availability of technical options to stop the disclosure of private information, which could potentially cause privacy violations. Since most social network users are in the habit of sharing pictures with other people online, the privacy of those in the photographs may be unwittingly compromised. This phenomenon is described as IPCPD. Their findings reveal that decisional control is generally a vital privacy protection tool in online social networks. Moreover, the importance of decisional control stems from different contextual situations specified by the "what" and "whom" aspects of information privacy.

Liang et al. [17] discussed another type of privacy concern online called deletion delay of photo sharing. They explored the possible access to a user's image even after deleting the image from social media platforms. They found that by using the Uniform Resource Locator (URL) of the image, it was possible to access the image anywhere from 7 to 30 days after the image was deleted. Popular social media platforms were also not immune to this problem. For example, on Facebook, it took up to 7 days for the image to entirely disappear. Also, it was observed that in cross-platform sharing, the original image from the source platform could still be accessed on the destination platform using the image URL on the destination platform.

Li et al. [18] investigated Amazon Wishlist and possible privacy exposures. They collected complete Amazon Wishlists of over 30,000 users and were able to make interesting observations based on the shopping preferences of users. To access the Wishlists, they constructed a crawler in Python (a programming language) that crawled through the search results for Amazon Wishlist search. They were able to predict shopping preferences based on gender, demographic groups, geolocation, and so on. Using machine learning and semantic analysis on Wishlist descriptions, they were able to extract users' private information. In their observation, they found that users tend to expose their activities, affiliations, educational backgrounds, and spouse names the most, thus compromising their privacy through the information provided about themselves.

Will et al. [19] proposed a system that would ensure that vendors would not be able to hold the personal information of users and store it for future use or sell it to other third-party vendors. In the proposed model, personal information is stored on the users' mobile device and requested by vendors when needed. In this centralized model, a relay service is used to hide data from vendors or websites, encrypt cache response, authorize vendors, filter unwanted requests, and provide features automatically like anonymous email. The authors proposed a model where personal information is stored on the users' mobile devices and requested by vendors when needed. Information can then be given in either a private or a trusted manner, and encrypted responses can be cached by a relay service. Vendors should only use the data inflight and never store personal information. This provides the user with data provenance and access control, while providing the vendor with accountability and enhanced security.

Harikant et al. [20] designed a study in which they modelled the behaviours of Facebook users based on their engagement with other users. They categorized these behaviours as *anomalous* and *non-anomalous*. If the users, based on their behavioural features, were showing *anomalous* behaviour, they were classified into different types of attacks invading the privacy of the said users. The behavioural features such as friend rate, comment rate, post-rate, and post-feedback rate determined the types of attacks on users. Based on some or all threshold values for these features, the attacks were categorized as compromised account attacks, sybil attacks, software attacks, identity clone attacks, creepers attacks, cyberbullying attacks, and clickjacking attacks.

1.2.3 Cybersecurity Awareness among College Students and Faculty

To investigate student and faculty members' awareness and attitudes towards cybersecurity, students and faculties in public universities in the San Francisco Bay Area of California were surveyed. The Bay Area is recognized for its most advanced community regarding wealth, technology, progress, and the diversity of the population [21]. For example, according to the San Jose State University website, 51% of its students are male and 49% are female. The diversity of students by ethnicity is 41% Asian, 26% Hispanic, 19% white, and 14% other. The average age of undergraduate students in fall 2017 was 22.6 years [22].

As part of this investigation, several surveys were administered. The first survey included ten general questions about cybersecurity awareness. The objectives were to understand students' awareness in such a tech-savvy environment of cyber-attacks (Silicon Valley) and to explore how they protected themselves against cyber-attacks. It is important to underline that the results of this study which are reported in this book are to show trends and cannot be generalized.

1.3 CONCLUSION

The review of the literature points to a few trends in student awareness in cybersecurity is as follows:

- Students do not have the requisite knowledge and understanding of the importance of information security principles and their practical applications.
- Security awareness is not due to a lack of security knowledge but in the way that students apply that knowledge in real-world situations.
- Students have a lack of knowledge of privacy policy and governing on the online social networks they used.
- Users tend to expose their activities, affiliations, educational backgrounds, and spouse names the most, thus compromising their privacy.

Despite the findings from the previous studies, we can observe that the diversity of the demographics and geographical location of the student population limits the generalization of the results. The studies do not sufficiently illustrate what students know/don't know and what behaviours they have/don't have to protect themselves in cyberspace.

In Chapter 2, the general survey results will be presented and discussed.

Students Cybersecurity Awareness

2

2.1 ONLINE CYBERSECURITY SURVEY

An online survey was designed to collect data about students' general awareness of cybersecurity and their overall behaviour and perception of privacy, authentication, and trust. Ten questions were constructed to measure the three dimensions of knowledge (what you know), attitude (what you think), and behaviour (what you do) [23]. The decision was made to keep the survey brief to increase the return rate. The study included the following ten questions:

- Do you consider yourself knowledgeable about the concept of cybersecurity?
- When using the computer system and Internet, what do you consider as being private information?
- On a scale of 1–10 (1 being the least secure and 10 being the most secure), rank how secure you think your communications are on each of the following platforms.
- Do you use a harder-to-guess password to access your bank account than to access your social networking accounts?
- Do you know what two-factor authentication (2FA) is and do you use it?
- Have you ever rejected a mobile app request for accessing your contacts, camera, or location?
- Do you ever reject app permissions?
- Do you have reason to believe that you are being observed online without your consent?
- Do you think that your data on the university system is secure?
- Do you think your communication through the learning management system (LMS) is secure?

The Qualtrics online survey software was used to deploy the survey. The online survey has been used in many investigations due to its effectiveness in reaching the targeted population, easy access and completion, rapid data collection, and cost of implementation. In spite of some disadvantages of survey studies (e.g., the respondents might report their desired behaviour [it is good practice to have a hard-to-guess password] rather than their actual real behaviour [saying that they have a hard-to-guess password]), these are the most cost-effective strategy to gather quantitative information.

The survey was sent to all (approximately 600) full-time and part-time students enrolled in different disciplines at California public universities in Silicon Valley in 2018.

The online survey was completed by 367 students (return rate of 61%). No demographic or personal identification data, besides gender and age range, was collected. Thirty-four per cent (124) of respondents were female, and 66% (243) were male. Fifty-four per cent (199) of the respondents were 18–24 years old, 42% (153) were 25–34 years old, and 4% (14) were over 35 years old. Thus, overall, the respondents were young, as expected for college students.

2.2 RESULTS

2.2.1 Knowledge of Cybersecurity

In response to the question regarding students' knowledge of the concept of cybersecurity, only 32% agreed that they are knowledgeable (agree or strongly agree), 47% believed that they have average knowledge (somewhat agree or neither agree nor disagree), and 21% reported that they are not knowledgeable (somewhat disagree or strongly disagree). The responses to other questions confirmed this self-evaluation. One-third of respondents do not have much knowledge of cybersecurity. Considering the respectively young age of the students, one can assume that this lack of knowledge is likely to be much higher in the general population, which is older. The differences between males and females and between age groups were not significant (1%–3%).

Seventy-nine per cent of respondents agreed that they are knowledgeable of cybersecurity. Overall, this can be considered a good percentage. However, the remaining 21% who do not know about cybersecurity is significant since most are younger college students assumed to have more knowledge of computers. It is also important to underline the considerable percentage (21% somewhat disagree and disagree) of participants who recognize not having any knowledge of cybersecurity. Now the question is, if they are knowledgeable, do they apply their knowledge for better cybersecurity?

It's important to note that the degree of awareness of students about these topics is still considerably better than the results of employees of companies. Mediapro (2017) surveyed more than 1,000 employees across the United States and found that seven of ten employees lack the awareness to stop preventable cybersecurity incidents [24].

Previous survey results from Slusky and Partow-Navid [10] also suggest that the major problem with security awareness is not due to a lack of security knowledge, but to how students apply that knowledge in real-world situations.

It is important to recognize that the survey questions were not very specific. The intention was to see, all things considered, how the students self-evaluate their knowledge of cybersecurity.

2.2.2 Privacy

The respondents were asked to select the information type that they consider to be private. The data that respondents found most private was bank account information (97%), contact information (83%), pictures (74%), location (73%), online search history (68%), internet protocol (IP) address on the device (64%), and others (18%). The students did not consider health information private, and because they are generally in good health, they did not have health records. There were no significant differences between female and male respondents and age groups. Overall, it seems that users consider their bank account, contact information, and location as the most private information.

The respondents were asked to rate the security of each platform on a scale of 1–10 (1 being the least secure and 10 being the safest). In this case, what was ranked (strongly agree and agree) higher were as follows: online banking (67%), mobile banking (61%), email (37%), texting and mobile texting (28%), and social media (15%).

The data overall suggests that

- Most people agreed that mobile banking is secure with 85% agreement. Only 10% disagreed or strongly disagreed overall with 5% being neutral.
- Respondents agreed that online shopping is mostly secure (63%), while 24% disagreed and 13% were neutral.
- People are somewhat positive about security in online shopping (32% somewhat agree).
- Social media is considered least private with a considerable disagreement of 49% and low agreement of 35%. Neutral reviews are also high at 16%.
- Forty-eight per cent of respondents agreed that texting is secure. People have a neutral stance towards texting (20% neither agree nor disagree).
- Messaging apps are similar to texting, where 51% agree they are secure and 31% disagree. Neutral reviews are 18%.
- Sixty-two per cent of people agree that email is secure, and 24% disagree. Neutral reviews are 13%.

For the question of whether or not respondents had ever rejected a mobile app request for accessing their contacts, camera, or location, 50% responded "Yes" (47% females and 51% males, 49% of the 18–24 age group and 48% of the 25–34 age group). The "No" responses were 7%, and 43% replied, "Not sure." The younger age group (18–24) rejected fewer app permissions to use contacts (7% "No") versus the older age group (25–34) of which 8% replied "No." Considering that 7% answered "No" to the question, "Do you ever reject app permissions?" and 43% said they sometimes rejected app permissions, we might extrapolate that there might be fewer people that reject app requests to access their unneeded data on a mobile phone.

How users manage app permissions is a good indicator of cybersecurity awareness and security in mobile and computing behaviour. App permissions are when users agree to give a mobile application permission to perform its full range of functions. For example, a mobile application might want access to various functions on users' smartphones and tablets, such as location and contacts. But some of these app permissions should not be granted. Why would a flashlight app need access to a camera or contact list? Negligence in giving access to what an app might request access for can endanger users.

The same type of issue is revealed for respondents who consider their bank account, contact information, and pictures as their most private information. In general, respondents considered almost all of the above information as private. However, they regarded their IP address, locations, and search results to be less private than their contact information, even though we know that most contact information is available on the Internet. A great deal of contact information can be purchased on the Internet for a few dollars from legal sources while IP addresses and location data might be harder to obtain. Hackers who obtain a person's IP address can then acquire precious information about users through their device, including their device's city, state, and Zone Improvement Plan code (a postal code used by the United States Postal Service). With this location data, hackers can find out other personal information about their potential targets.

Interestingly, when respondents were asked if they have rejected app permissions, 50% said "Yes," 43% "Sometimes," and 7% "No." The result might indicate that when it is easy to see which application asked for users' permission; users might make a judgement not to let the application access data they consider being private. This result confirmed a previous survey [25], which reported that 92% (153 participants) of those surveyed expressed that they have rejected access if they believe the app does not need to access the camera or contacts. This result is also in line with a previous study by Haggerty et al. [26] who found that 74.1% of iOS (a mobile operating system created and developed by Apple Inc.) users would reject the app permissions list. Thus, even expert users may grant information access too light-heartedly. However, in many instances, users do accept granting permissions requested by the majority of applications, and the percentage in this study is much lower than in the previous survey.

Another parameter that still illustrates awareness of cybersecurity among college students is the use of 2FA. Around 80% claim that they use it. However, only 28% use it always and 54% for some accounts. Still, about 20% don't use or do not even know what it is.

Even the usage of 2FA among respondents who self-evaluated as knowledgeable (strongly agree, agree, or somewhat agree) is not much different. Thirty-one per cent say they always use 2FA, 54% for some accounts, and 15% said they do not know what it is or do not use it at all.

While we see a low employment of preventive measures being taken by college students, it is interesting to observe that 60% of respondents report that they have reason to believe that they are being watched online without their consent. This percentage is 51% among the people who self-evaluate as knowledgeable about cybersecurity (strongly agree, agree, or somewhat agree). It is also important to underline that despite awareness of the importance of Internet privacy, college students are still willing to engage in risky online activities [27]. Consequently, young adults need to be motivated to enact security precautions. They are advised to take the risks of Internet use or online safety communication seriously and consider it as personal responsibility [28].

2.2.3 Passwords

Participants were asked if they used a harder-to-guess password to access their bank account than to access their social networking accounts. Fifty-eight per cent stated that they use a harder-to-guess password for their bank account than for social networking. Seventeen per cent used harder-to-guess passwords for their bank account, 25% said they use the same passwords for both, and 58% use different passwords but the same password complexity for both. In this case, the difference between females and males is 7%, and the difference between the two age groups (18–24 and 25–34) is 3% for "No, use the same for both." However, for "No, use different for each but the similar level of complexity," the difference between females and males is 1.5% and between age groups is 9%.

Overall, the data suggests that

- Most users report that they use more difficult passwords for their bank accounts compared to other passwords. More than 57% of the overall participants responded to having a more complex password for bank accounts.
- Male users had more "yes" answers than female respondents, while men had fewer "No, use the same for both" and slightly more "No, use different for each but a similar level of complexity" answers.
- Age can have an impact on making such decisions like making the banking password more complicated than those being used in social media accounts. Older aged people tend to make their passwords more secure.
- Among the female population, 54% have a more difficult password for bank accounts, while among the male population, 60% use a more difficult password.
- In the 25–34 age group, there is a slightly larger number of participants (61%) who use a more difficult password for banking accounts.

While one might consider that a hard-to-guess password is only essential for sites that save private information, an easy-to-hack social networking account can open the door for a lot of social engineering or ransom attacks. Therefore, a hard-to-guess password is needed for all types of accounts that include user data.

This issue also indicates that 54% of respondents use 2FA for some accounts and 27% use it for all accounts, 11% do not use it at all, and 6% do not even know what it is. Considering that the respondents in this phase were university students, this seems to be a very alarming issue. It is also an indicator that cybersecurity knowledge does not routinely transfer in every case to more secure behaviour.

It seems that among college students, adopting and using better security practices still needs to be improved.

Fifty-four per cent of respondents (60% females and 50% males and 50% of 18–24 and 57% of 25–34 age groups) use 2FAs for some accounts, and 27% (15% females and 34% males) use for all accounts.

2.2.4 Trust

Sixty per cent of respondents had reason to believe that they were being observed online without their consent, 17% had no reason to believe, and 23% were not sure. The perception of being seen might be an indicator of the trust of Internet privacy.

Another important factor in security is the perception of trust in computer systems. This was evaluated through three questions: (1) "Do you have reason to believe that you are being observed online without your consent?" (trust of the Internet), (2) "Do you think that your data on the university system is secure?" (trust of the university system), and (3) "Do you think your communication through LMS is secure?" (trust of the LMS).

Interestingly, 60% of respondents (60% females, 60% males, 58% aged 18–24, and 63% aged 25–34) believe they are observed online without their consent. It seems that the percentage goes up with older age groups. It would be interesting to investigate what factors make them believe they are watched online and how. Is it just their search behaviour, or more? Moreover, if the respondents feel that they are observed online, then how do they think they are observed? Also, what, if anything, do they do about it?

2.2.5 Trust of University Data Security

For the perception of data security of university systems, only 10% believed that their data was secure (8% females and 11% males), 56% believed that data was relatively secure on university systems (63% females and 52% males), 21% declared not secure (16% females and 23% males), and 14% declared not sure.

Overall, the data suggests that

- Despite a lower number of female respondents, only 8% tend to think the university system is fully secure.
- Fifty-six per cent (204) of the respondents think that the system is relatively secure.
- Only 21% (77) think that the system is not secure out of the 367 respondents.
- The results in the different age groups are the same although the numbers of respondents in the age group of 18–24 are more than those in the age group of 25–34.
- The result shows that most of the people chose the option, "Relatively secure," which explains their insecurity when it comes to the security of their data in the university system.
- Forty-five per cent think that data on the university system is not secure or they are not sure.
- Only 10% of the respondents believe the application is secure.
- Sixty-nine per cent (106) of the respondents aged 25–34 lean more towards believing the application is secure or relatively secure.
- Respondents aged 18–24 are more "unsure" about the security of their data on university systems than the older respondents.
- Eleven per cent (26) of males versus 8% (16) of females think that the university system is secure.

For the last question concerning the LMS, 9% believed it to be secure, 43% relatively secure, 17% not secure, and 31% not sure.

Overall, the data suggests the following (Table 2.1):

- The percentage of the respondents believing that the LMS is completely secure was a meagre 9% (33).
- The ratio is almost similar among all age groups and between genders. Consequently, we can assume that the trust level of LMS being completely secure is doubtful. Since 52% agreed that it's "Relatively secure," this

TABLE 2.1 General students' survey results

AWARENESS PARAMETER	PERCENTAGE OF RESPONSES	PERCENTAGE OF RESPONSES CHARTS
Knowledgeable about the concept of cybersecurity	33% agree or strongly agree 47% average knowledge, somewhat agree or neither agree nor disagree	7% / 13% / 13% / 34% / 26% / 6%
Consider being private information	97% bank account information 83% contact information 74% pictures 73% location 68% online search activity 64% IP address on the device 18% others	97 / 83 / 74 / 73 / 68 / 64 / 18
Security of each platform (strongly agree and agree)	67% online banking 61% mobile banking 37% email 31% online shopping 29% mobile texting 28% texting 15% social media	67% / 61% / 37% / 31% / 29% / 28% / 15%
Harder-to-guess password to access your bank account than to access your social networking accounts	58% harder-to-guess password for bank account than for social networking 25% different passwords, but the same complexity for both 17% same complexity for both passwords	58% / 25% / 17%

(*Continued*)

TABLE 2.1 (*Continued*)   General students' survey results

AWARENESS PARAMETER	PERCENTAGE OF RESPONSES	PERCENTAGE OF RESPONSES CHARTS
2FA usage	54% 2FAs for some accounts 28% for all accounts 11% do not use it 6% No, don't know what it is	54% 28% 11% 6%
Mobile app permission request	50% "Yes" 43% "Sometimes" 7% "No"	50% 43% 7%
Do you have reason to believe that you are being observed online without your consent?	60% believe that they were being observed online without their consent 23% were not sure 17% had no reason to believe	60% 23% 17%
Security of data on the university system	10% believed that their data is secure 56% relatively secure 21% not secure 14% not sure	56% 21% 14% 10%
Security of data the LMS is secure	9% believed secure 43% relatively secure 17% not secure 30% not sure	43% 31% 17% 9%

shows that people do have some faith in the communication system but suspect there might be some loopholes which could be exploited to make the system vulnerable.

- Seventeen per cent of the respondents believe that the communications are not at all secure.
- Almost 30% of people are not sure about their communications being secure. They are probably not aware of the security features currently being used in the system.
- Less than 10% of the total number of people who took the survey believed that their data is secure on the university system.
- Forty-three per cent of people who took the survey are more inclined to believe their data is relatively secure on the system.
- Almost one-third of people who took the survey are not sure of whether they should trust the system or not.
- The percentage of people older than 25 who have more faith in the system is 14.38% (22).

The level of trust in the security of the university system (including LMS) is not very high. Only 9% (11% of female respondents and 8% of males) think that the system is secure. However, 43% consider it to be relatively secure. It is also important to underline that 17% believe it is not secure. Overall, we can observe that 50% of the students overall trust the university data management system (Table 2.1).

2.3 CONCLUSION

Overall, 79% of the respondents self-evaluated their cybersecurity awareness as average or more. Sixty per cent believed that they are being observed online without their consent. However, the data does not indicate that the knowledge they claim to have translates to secure behaviour or that the awareness that they have is very practical.

For example, they do not consider IP address or location as very private, or they have a hard-to-guess password for their bank account (which is a good habit), but they do not apply hard-to-guess passwords for all of their accounts. The respondents easily grant permission to apps to access their contacts, camera, or location (Table 2.1).

Interestingly, the students do have a measured level of confidence in the security of the university system and LMSs since 56% (204) think that the system is relatively secure. Now, the questions that we might ask are as follows:

- Do the students protect their privacy?
- The students rate the social networking lowest on security. What would be their behaviour in using social networking?

Sixty per cent of students believe that they are being observed online without their consent. However, that does not change their online usage behaviour. What would explain this behaviour? How is this related to their passivity vis-à-vis security issues?

Follow-up questions could be

- Do they set their privacy level on their favourite browser or at least review it?
- Do they know what cookies are?
- Do they delete cookies?

In Chapter 3, we will explore the responses to some of those questions with several additional follow-up surveys.

Students' Secured and Unsecured Behaviours

3

Follow-Up Studies

3.1 INTRODUCTION

The results of the first general survey study of overall cybersecurity awareness led to several follow-up survey studies. These studies were conducted to collect more information about students' awareness of privacy, two-factor authentication (2FA), identity theft, ransomware, passwords, mobile phone protection, Internet cookies, and social media. The objectives were to better understand the students' behaviour when using social media, and to know how they protect themselves in cyberspace. Three separate surveys were conducted. The questions in each survey were grouped to cover the three topics: identity, trust, and privacy. The surveys included questions in the following areas:

- Privacy
- 2FA
- Passwords
- Identity theft
- Ransomware
- Mobile phones
- Internet cookies
- Social media

In the following sections, the results of the follow-up surveys are summarized for each area in each group of questions. In all the surveys, no identifiable information about the participants was collected.

3.2 RESULTS

3.2.1 Demographic

Privacy survey respondents

Eighty-nine students completed the online survey. Forty per cent (35) of respondents were female, and 60% (53) were male. Fifty-one per cent (45) of respondents were 18–24 years old, 43% (38) were 25–34 years old, and 6% (5) were over 35 years old. Forty-four per cent (39) were undergraduate students, and 55% (49) were graduate students.

Identity survey respondents

One hundred and seven students completed the online survey. Thirty-nine per cent (42) of respondents were female, and 61% (65) were male. Fifty-two per cent (56) of respondents were 18–24 years old, 42% (45) were 25–34 years old, and 7% (5) were over 35 years old. Forty-six per cent (50) were undergraduate students, and 54% (58) were graduate students.

Social networking respondents

Eighty-three students completed the online survey. Thirty-seven per cent (31) of respondents were female, and 63% (52) were male. Fifty-three per cent (43) of respondents were 18–24 years old, 44% (36) were 25–34 years old, and 4% (3) were over 35 years old. Forty-eight per cent (40) were undergraduate students, and 52% (43) were graduate students.

Although the number of respondents in each group was somewhat different, the compositions were almost identical. Thus, it is realistic to assume that most of them are majoring in computer engineering. The intention was to capture those students that might be more knowledgeable in cybersecurity and have computer expertise.

3.3 PRIVACY

One of the main characteristics of the information age is information sharing. People use multiple devices to copiously share information about their lives, activities, and their opinions and beliefs, just to name a few. It is the ease of this connectivity that has brought about a sense of a virtual community that presents itself in the form of social media and social networking. According to Google, 80% of people use a smartphone every day in the United States. On average and at any given hour, 75% of users are using social media apps such as Facebook and Twitter. The use of social media apps peaks during rush hour and night-time, which leads to the assumption that social media is often used to fill idle time, similar to socializing with friends after work or school but without the personal interaction [29].

An article by CNN (Cable News Network) states that "Americans devote more than 10 hours a day to screen time" and that increasing screen time is taking away from other valuable day-to-day practices [30]. This data supports the assumption that the combination of constant information gathering on the web and the use of social media when extra time is available results in nonstop information sharing over the Internet.

There is a lot of ambiguity in what information is actually being shared and with whom, on social media platforms. In addition, it is not clear if users have a good understanding of the importance of Internet privacy. Do users understand that they are sharing sensitive data? Do users understand what it means when their personal information has been compromised? Do users understand the importance of Internet privacy? One of the goals of the follow-up study was to analyse the data extracted from questions related to privacy.

The following questions were asked in regard to privacy:

- Do you participate in online surveys, promotions in hopes of receiving goodies, gift cards, or prizes?
- Which of the following permissions are you most likely to give to third-party apps services online?
- Do you grant all permissions to every application while installing?
- Which of the following personal information are you willing to submit to applications/services that ask for such information?
- Is Google location service and/or other location services active on your phone (global positioning system [GPS] enabled)?
- On social media platforms like Facebook, do you give permissions to entertainment-based apps like "Which Game of Thrones character are you"?
- How likely are you to accept app/game invites on Facebook?
- Do you give permission to use your location to apps on your phone?
- What are the privacy settings on your favourite browser?
- Do you know what cookies are on the Internet and what they do?
- Do you accept cookies when browsing on the Internet?

3.3.1 Results and Explanation

Fourteen per cent (13) of respondents participate in online survey promotions in hopes of receiving goodies, gift cards, or prizes; 36% sometimes participate; 15% only give limited personal information; and 34% do not participate at all.

Sixty per cent (76) of respondents grant permission to third-party apps services online to access their first and last names, 14% (18) grant permission to access their contacts and phone book, 18% (23) grant permission to access their location, 3% (4) grant permission to read their messages, and finally, 4% give all permissions.

Eleven per cent (10) declare that they grant all permissions to every application while installing, 47% (42) grant permissions which they deem necessary, 16% (14) grant all permissions to popular apps, and 27% (24) grant some permissions which they deem necessary to popular apps.

Eleven per cent (10) of respondents feel comfortable to submit their first name, last name, email, address, phone number, and other information like employment to applications/services.

Sixteen per cent (14) of respondents just share their first name, last name, email, address, and phone number. Twenty-two per cent (10) of respondents share their first name, last name, email, and phone number only; 43% (39) of respondents share their first name, last name, and email; and only 8% (7) do not give any information at all.

Twenty-three per cent (21) have their Google location service and/or other location services active on their phone (GPS enabled) all the time. Seventy per cent (63) have their Google location service and/or other location services sometimes on while using maps or ride-sharing services like Uber. Seven per cent (6) always have location service off.

Eighteen per cent (16) of respondents on social media platforms like Facebook give permissions to entertainment-based apps like "Which Game of Thrones character you are" because they think it is fun, whereas 82% (74) don't give permissions.

Four per cent (4) generally accept app/game invites on Facebook to check out the deal, 19% (17) accept only sometimes, and 77% (69) never accept a game invitation.

Ten per cent (9) of the participants give all apps permissions to use their location on their phone, 31% (28) give permissions only to the apps that they trust, 48% (43) give permissions only for necessary services like ride sharing, and 11% (10) don't give permission at all.

Twenty-five per cent (22) of respondents leave the default settings on their favourite browser without checking, and 10% (9) review the default settings but do not change them. Forty-two per cent (37) change the privacy settings to give themselves more privacy. Seven per cent (6) change the settings for less privacy. Seventeen per cent (15) don't remember what they did.

Thirty-two per cent (29) of respondents know what cookies on the Internet are and completely understand how they work. Forty-three per cent (39) say they have some understanding of how cookies work. Seventeen per cent (15) have heard about cookies but do not understand how they work. Finally, 8% don't know anything about cookies.

Nineteen per cent (17) of respondents accept all cookies, 54% (48) accept selected cookies, 15% (13) do not accept any cookies, and 12% (11) don't know about cookies.

The results of this follow-up survey study provided some useful insights into the privacy concerns among college students. The survey helped the researcher precisely pinpoint the areas where privacy concerns are high, or the students who are at more risk of getting their privacy violated. Overall, awareness among students concerning their privacy online was relatively high, which is a good sign.

Consider that

Forty-nine per cent give limited personal information or do not participate in an online survey to receive goodies.
Sixty per cent give very little information about themselves to third-party apps.
Forty-seven per cent grant permission to apps which they deem necessary.
Seventy-seven per cent never accept a game invitation.

Seventy-nine per cent give permission to locations when needed.
Fifty-two per cent check the privacy settings on their browser and might change them.
Thirty-two per cent know what cookies are.
Fifty-four per cent accept selected cookies.

Given the results of the privacy questions, we can conclude that over 50% of students have some awareness about their privacy. However, the results find that a large number of students still show risky behaviours online. For example, they show insecure behaviour when they give permission to apps to use their location, even when not needed.

Responding to online surveys from unknown entities can also be a risky behaviour. Criminals might use this type of approach to collect users' private information or might include clicks to a phishing link.

Location services in smartphones and in-car navigation systems give users access to many useful services, for example, finding the nearest gas station, but there are serious issues about how vendors who gather that data use and sell it, most often without users knowing. The location data may be shared with third parties unknown to the users. Because users do not know who the third parties are or how they might use people's data or whether those entities are trustworthy or not, it is possible that users' information may be compromised. Location data can also be used to track the pattern of users' behaviour. It can also be used to steal the identities of users when the location data is disclosed, particularly when the data is combined with other personal information. Location data can also be used to track the records of an individual's movements and activities.

Not knowing about the roles of cookies or privacy settings is another area of concern. Cookies might contain a lot of information. If they are not encrypted and protected, they will be readily available to anyone who may compromise users' systems, such as sensor details (a device, module, or subsystem that detects events or changes in its environment and sends the information to other electronics). A sensor is always used with other electronics from people's mobile devices. Thus, students are much less cautious about protecting themselves from the dangers of social hacking, relating specifically to sharing information on social media. They need to avoid participating in online surveys or giving out information for rewards. Consequently, they may face risk and might not take strict measures to protect their privacy based on their personal views of Internet privacy.

3.4 TWO-FACTOR AUTHENTICATION

Soon after Google first introduced its 2FA system, many vendors like Twitter, Facebook, and Dropbox also followed Google's lead. Usually, 2FA is not enabled by default. Users are encouraged to manually allow the feature for improved protection.

A typical behaviour among users is to write or save their passwords or personal identification number (PIN) for easy remembering. For example, in one survey, it was reported [31] that 50% of the respondents wrote down their passwords. Writing passwords is often tied to the password policies requiring employees to change their passwords periodically, making passwords harder to remember. Another study by Gunson et al. [32] showed that many users remember a few passwords and reuse them for their online accounts. A single password is reused at least twice. However, the magnitude of password reuse increased over time as users gained more accounts but did not create more passwords. Other studies show that typical users have 10–50 online accounts where they need passwords. These users only had one to seven unique passwords, which they reused repeatedly.

According to Adams and Sasse [31], users also tend to choose easily memorable passwords. They tend to use their date of birth or loved ones' information as passwords. Users consider PINs to be more difficult to remember than passwords. These issues can be addressed by setting stricter password policies, which should improve users' password choices. Users should be educated and made aware of the implications of using a weak password.

2FA involves more than one type of authentication. These authentication components are classified into three categories:

1. What you know—information that is only known to the individual, for example, password
2. What you have—a unique physical token, for example, automated teller machine card
3. What you are (inherent)—the intrinsic property of the individual (e.g., fingerprint)

In the follow-up survey, the following questions were asked regarding 2FA:

- Are you aware of 2FA and its benefits?
- Do you use 2FA in your online accounts?

3.4.1 Results and Explanation

The same type of 2FA questions that was asked in the general survey (Chapter 1) was also added in the follow-up survey, but in a different context and with a different group of respondents. The results show that

Eighty-three per cent (90) of respondents are aware of 2FA and its benefits, 14% (15) are not aware, and 3% (3) are not sure.

Twenty-five per cent (30) of respondents use 2FA on all their accounts when available, and 34% (41) use it on some of their accounts when available. Nineteen per cent (23) apply it only when it is required. Twelve per cent (14) do not use 2FA, and 10% (12) do not know what it is.

These results are in line with the first survey results (Chapter 2) that showed 54% of respondents use 2FA for some accounts, 27% use it for all accounts, 11% do not use it at all, and 6% do not know even what it is.

In the follow-up survey, 83% are aware of 2FA and its benefits, and 59% use 2FA on all or some of their accounts. However, 33% do not know or use it unless it is required.

2FA is one of the easiest ways to prevent hackers from hijacking people's accounts. Even when the users have a very strong password, when database breaches happen, hackers can copy and paste passwords. Thus, the 2FA is a good second line of defence.

3.5 IDENTITY THEFT

Identity theft is a way someone tries to steal information, which resembles or identifies the information of someone else. It is described as an unending crime in this information age. The information includes a person's birth certificate, social security number, and driver licence. Hundreds of millions of identities are exposed every year; tens of millions of these identities are exploited further. After the retrieval of the information, the thieves use the information to initiate credit card accounts, apply for jobs, apply for loans, and for other fraudulent works in the victim's name. Most of the time the thieves use the victim's information to buy merchandise and other goods by using already-established accounts.

The consequences of identity theft can be devastating such as leaving the victim with a significant amount of debt and damaging their credit card rating. They can also be denied loans for a car or a house due to their bad credit card history.

Identity thieves often target school and college-age students whose credit card scores are clean. The young students are primary targets because they have to fill many applications and forms for colleges and for other uses like credit card offers, online purchases, and financial assistance applications, and for pro versions of Google Play and iOS applications. Students are often unaware of their victimization. Unfortunately, this also makes students more susceptible to be the first targets for identity scams and fraud.

According to the Federal Trade Commission's (FTC) 2016 Consumer Sentinel Network report, 19% of identity theft complainants were under 29 years old. In 2016, 74,400 young adults fell victim to identity theft [33].

Filling out forms and data sharing on school and college campuses have turned out to be second nature to numerous students who have access to cell phones, broadly accessible Internet, and social media websites. Many college students get emails with pre-approved credit offers. The mass mail which the student receives is pre-filled with student's information like name, address, and another victim's details. If the student is not interested in the offer, they might discard the mail and throw it in the trash. Obtaining the pre-filled form is one common way thieves can get access to the personal information of students. The thief can fill out the rest of the form and send it, which allows them to gain access to the victim's identity.

Another way identity theft occurs is when thieves get access to the personal banking information of the victim using checking or savings account statements. Any student who is not keeping track of their financial statements regularly is susceptible to fraudulent charges. Thieves might withdraw money in small amounts, which are not enough to be noticed and thus are considered to be errors.

The follow-up survey questions regarding identity theft and protection included the following questions. The purpose was to investigate how many students are aware of the concept of identity theft and how they protect their identity.

- How often do you give your social security number while applying for a job?
- How often do you check your financial statements to ensure there is no suspicious activity before paying the credit card bill?
- How often do you log out of your account from shared devices?

3.5.1 Results and Explanation

Twenty-five per cent (27) of respondents always share their social security number while applying for a job, 20% (22) share most of the time, 3% (3) share half of the time, 17% (18) share sometimes, and 36% (37) never share their social security number.

Nineteen per cent (20) of respondents often check their financial statements to ensure there is no suspicious activity before credit card bill payment, 10% (11) never review their statement, 28% (30) check most of the time, 35% (37) check always, and 9% (9) check half of the time.

Sixty-five per cent declare they always log out of their account from shared devices, 14% (15) about half of the time, 11% (12) most of the time, and 10% (11) never.

The data on identity protection is more alarming considering students are frequent victims of identity theft. According to Experian [34], there were 13,852 identity theft complaints to the FTC in 2017 affecting children and teens (age 19 and under), which represents 3.89% of all identity theft complaints for the year.

Since students are often searching for jobs, they frequently submit job applications. As such, they are at high risk of being exposed to fake job postings and social engineering hackers. Forty-five per cent of those who share their social security number most of the time do so without cross-checking to see if the company is legitimate, and this is hazardous behaviour. According to some estimates, hackers have stolen 60%–80% of social security numbers [35].

Generally, it seems that students are more aware of checking their financial statements and logging out from their accounts on shared devices. That said, still 19% check financial statement half of the time and 35% never or seldom indicate a very high risk factor.

Twenty-five per cent of students share their social security number while applying for a job, and 20% share their social security number most of the time. This percentage is very significant and illustrates the lack of awareness students have of the importance of their information.

3.6 RANSOMWARE

Ransomware is malware that stealthily gets installed on users' devices and holds their files or operating system functions for ransom. It restricts the user from using their PCs (personal computer) or mobile devices by encrypting their files. Paying ransom (through Bitcoin, a cryptocurrency, a form of electronic cash), however, does not guarantee that the victim will get their files back. In the most recent statistics, 184 million ransomware attacks were carried out in 2017 in the United States. It represents a 71.2 drop growth from 2016 [36].

Ransomware falls into two types: basic ransomware and CryptoLocker ransomware. In basic ransomware, the computer screen freezes and the victim is asked to pay the ransom. In CryptoLocker ransomware, the computer and software installed in the system work fine, but the user's files get encrypted. The attacker retains the only decryption key stored on their server. It becomes impossible for the victim to unlock their files without the attacker's assistance.

Ransomware can be downloaded onto systems when users unknowingly visit malicious or compromised websites. Some ransomware is delivered as attachments from spammed emails or downloaded from malicious sites or by exploit kits (automated threats that utilize compromised websites to divert web traffic, scan for vulnerable browser-based applications) and run malware onto vulnerable systems. After the ransomware is downloaded onto the system, the personal files in the computer get encrypted. Then, a ransom message is displayed, stating the deadline and amount of ransom demanded by the attackers in the form of Bitcoin. Victims then have to use a TOR browser (which stands for "the onion routing," which is a web browser designed for anonymous web surfing and protection against traffic analysis) to pay the ransom. In 2017, most private and public sectors were infected by ransomware, which exploited thousands of users around the world and demanded Bitcoin payments.

Young people and students should be aware of this threat and should know how to prevent it from entering their systems. It is better to be careful than to pay ransoms to get the private data back. Also, it is not guaranteed that after a victim pays the ransom amount, he/she will get back their documents. Attackers may demand more money to give the data back.

The following questions were asked to understand the awareness of ransomware in the identity follow-up survey:

- Have you or your peers/friends fallen victims to ransomware in the past?
- Do you know what you should do if infected by ransomware?
- What would be the cost you would be willing to pay if your computer got infected with ransomware tomorrow?
- Do you make regular backups to avoid paying ransom if your computer is affected by ransomware?

3.6.1 Results and Explanation

Sixty-five per cent of respondents have never seen anyone infected by ransomware, 6% have had their systems affected by ransomware, and the remaining 29% have had their peers/friends face attacks.

Fifty-six per cent (59) of respondents have never seen anyone infected by ransomware, 33% (35) have seen peers/friends fall victim to ransomware in the past, and 11% (12) have had their own computers infected by ransomware.

Forty per cent (41) have no idea what they should do if infected by ransomware, 4% say they would pay the ransom and hope that the attacker gives back the access, 21% (22) will take a photograph of the ransom note and file a police complaint with cybercell (the Federal Bureau of Investigation [FBI] unit in charge of investigating cyber-attacks by criminals, overseas adversaries, and terrorists), and 36% (38) say they will use anti-virus malware to clean the ransomware from the computer.

Fifty-four per cent (57) of respondents say they wouldn't pay any money if their computer got infected by ransomware tomorrow, 24% (25) would accept to pay under $100, 13% (14) would pay between $100 and $500, 6% (6) would pay between $500 and $600, and 3% would be ready to pay over $1,000.

Thirty-six per cent (38) of respondents make regular backups to avoid paying ransom if their computer is affected by ransomware, 33% (35) don't make any backups, and 31% (33) back up once in a while.

Forty-seven per cent (50) of respondents believe they might be targeted by ransomware in the future, 25% think they most likely will be a target of ransomware in the future, and 28% (30) think they will not be targeted by ransomware.

The number of ransomware attacks is growing. Interestingly, 35% of students surveyed said that they themselves or their peers had been affected by ransomware. Seventy-two per cent of respondents believed they maybe or most likely would be targeted by ransomware in the future. Thus, a large number consider ransomware as a considerable risk. That said, 40% of respondents would not have an idea what they would do if infected by ransomware.

Only 36% of the surveyed students back up their systems, which means that almost two-thirds of students do not make regular data backups. In the case of a ransomware attack, they would be left with no choice but to pay the ransom and expect to regain access to their system or lose access without spending the money.

Fifty-four per cent of respondents say they would not pay ransom, but 46% might pay based on the amount reasonable to them. It should be known that only half of the respondents who paid the money after being attacked by ransomware were given back access to their files/system. As such, it's important to recognize that paying ransom can validate attackers (since they got what they asked for) and could lead to more attacks in the future.

3.7 PASSWORDS

From the earliest days of web applications, the username and password-based mode of authentication has been the most popular way of verifying user access. There is a multitude of results for the search query "password hacking apps" on Google or any other search engine provider. Password hacking remains one of the significant security threats to the digital world, even after so much technological advancement.

Although many new ways of authentication have been introduced like Quick Response (QR) Code scanning and password-less logins using confirmation codes sent to phone numbers or email IDs, all the new authentication mechanisms require access to a device with previous setup ready. For the sake of simplicity, most people tend to use username and password-based mechanisms, even when other modes are available. With the growth of web applications and most tasks done online, an average user has so many online accounts that it is difficult to keep track of all of the usernames and passwords. The users tend to repeat their usernames and sometimes their passwords across accounts or they use simple passwords that are easy to remember. This general behaviour of people makes their accounts vulnerable, as easy-to-remember authentication credentials are also simple to crack or guess.

There are often jokes circulating among college students related to how the requirement of stronger passwords is exaggerated. This state of mind may sometimes undermine young people's thought-process about the stringent requirement for a stronger password for overall safety.

This part of the study aims to capture the behaviour of college students in choosing and maintaining their passwords. The intention is to observe patterns among young people who have either not faced implications of security hacks or do not fully understand the direct threat posed by the hacker community to their privacy, financial data, safety, and security.

In order to find out more information about students' behaviour and their usage of passwords, the following questions were asked in the identity surveys:

- Do you reuse the same password?
- How often do you change your password?
- Do you share passwords?

3.7.1 Results and Explanation

Thirty-nine per cent (41) of respondents say they reuse a password, 39% (40) say they sometimes reuse, and only 23% (24) say they do not reuse a password.

Fifty-one per cent (54) of respondents change their passwords only when it is required, 7% (7) after over 6 months, 29% (31) every 3–6 months, and 13% (14) in less than 3 months.

Eighty per cent (85) of respondents say they do not share passwords, 15% say they sometimes share passwords, and 7% say they do share passwords.

Seventy-eight per cent of respondents reuse or sometimes reuse their passwords. Reusing passwords might put users at risk when some websites do not encrypt username and passwords and are using plain texts. There may also be some security vulnerabilities such as SQL(Structured Query Language) Injection, OS (operating system) Injection, and others that again might expose user credentials to hackers. Once a hacker has the user's credentials, they can apply them to other websites such as iCloud (a cloud storage and cloud computing service from Apple Inc), banking sites, Facebook, Google, and Amazon. If users are using 2FA, they can still manage to circumvent the situation, but if 2FA is not used (and the later survey reveals most people are not using 2FA), then they are at total risk of compromising their data.

Fifty-one per cent change their passwords only if it's required and 22% share passwords. Regular password changes have been standard practice for a lot of reasons including password sharing. Despite the fact that people are instructed to never share their password, this practice is still common especially among students in order to view movies or share account info on Netflix, Amazon, and other similar sites. An instructor might share a password to give an assistant quick access to a file. Thus, despite the hardship of frequently changing passwords and the challenges of remembering them, changing passwords is considered a good security measure.

3.8 MOBILE PHONE PROTECTION

In the last few years, there has been a paradigm shift in computing devices and mobile devices are becoming more popular for everyday needs. Although there are many sophisticated security measures introduced in both iOS and Android devices, it is essential to secure the most vulnerable entry point in the security chain that is the user itself. Many users might think that no one is interested in stealing his or her data from mobile phones and often do not use a hard-to-guess and difficult PIN or pattern.

This part of the study was aimed at investigating student awareness of mobile phone security. The following questions were asked in the survey:

- What is your preferred mode of mobile phone authentication?
- How often do you update applications on your cell phone?
- Do you grant all permissions to every application while installing?
- Which of the following of your personal information are you willing to submit to applications/services that ask for such information?
- Is Google location service or other location services active on your phone (GPS enabled)?
- Do you give apps permission to use your location on your phone?

3.8.1 Results and Explanation

Forty-six per cent of respondents preferred biometrics as a mode of authentication, 36% (38) PIN, 9% (10) pattern recognition, 2% voice command, and 7% other types.

Thirty-one per cent (33) always update applications on their cell phone immediately when they are available, 34% (36) update once in a while, 8% (8) update rarely, and 28% (30) use automatic updates.

Eleven per cent (10) declare that they grant all permissions to every application while installing, 47% (42) grant permissions which they deem necessary, 16% (14) grant all permissions to favourite apps, and 27% (24) grant some permissions that they deem necessary to popular apps.

Twenty-three per cent (21) have their Google location service or other location services active on their device (GPS enabled) all the time. Seventy per cent (63) say they sometimes keep location services on, while using maps or ride-sharing services like Uber, and 7% (6) never keep location service on.

Ten per cent (9) of participants give all apps permission to use their location on their phone, 31% (28) give permission only to the apps that they trust, 48% (43) give permission just for necessary services like ride sharing, and 11% (10) don't give permission at all.

Updating mobile phone applications and the general software system with the newest version is a significant security measure. Fifty-nine per cent of the surveyed students update immediately or use the automatic update feature, and 41% update once in a while or rarely. Since security patches are often corrected and improved through these updates, not updating these applications is considered dangerous behaviour.

Respondents seem to be very careful about giving permission to app access requests, since 74% give permissions only to those that they deem necessary, and only 11% grant all permissions for location services. Students who do not pay attention to the permissions that are given to each app might put themselves at a very high risk. If a user approves one of the permissions, the app gets all of the permissions from the same group automatically, without additional confirmation. For example, if an app gets permission to read short message service (SMS) messages, then it will also be able to send SMS messages, read SMS messages, and perform other operations from this group. Alternatively, permitting camera access allows the app to use the user's phone to take photos and record videos and this could lead to secretly recording videos or pictures at any time. One often-ignored factor is when users allow apps to access their entire address book, as this is a frequent and attractive target for spammers and fraudsters. Allowing access to contacts might also grant access to the list of all of the accounts users have on the device such as Google, Facebook, and Instagram.

As discussed previously, giving access to location is a very risky behaviour. Once the app knows where you are at all times, it might, for example, let burglars know when you are far away from home.

3.9 INTERNET COOKIES

3.9.1 What Are Cookies?

HTTP (HyperText Transfer Protocol) cookies are the cookies generated and altered by the server, which are further stored by the browser but transmitted between browsers and servers at each interaction. The working of cookies is defined in the following steps:

- The client communicates with the web server for the first time. The server generates a response for the request where an ID (session identifier) is generated and included in the cookie.
- The server sends the cookie to the client as a part of the header that is then stored in the client's browser. Each time a new query to the server is issued, this cookie will interact with the server. When a user enters a UR (Uniform

Resource Locator) in the browser, a search is made by the browser in the local memory to check whether there is any cookie associated with it or not. If there is a cookie, then it is put along with the query and sent to the server.

- The next step is content retrieval. The web page may contain additional links such as images and videos that correspond to other servers. Thus, a single web page can include content from more than one server.

First-party cookies are the cookies that are associated with the web server indicated by the page URL that the client is visiting. The first-party server sets their first-party cookies, whereas advertising and marketing companies mostly use third-party cookies. Browsers can retrieve these cookies, while the user visits a web page that includes the content from third-party providers such as images, videos, and others.

Cookies contain attributes such as the domain, name, and expiration date of the website, and can also include the following data:

- Credentials such as names, passwords, and identifiers
- Session data and data from the website
- User tracking information

3.9.2 Cookies and Privacy

Cookies have been used for user identification and online advertising and marketing for a long time. However, consumer proponents, policymakers, and even advertisers themselves recognize the prospective danger to user privacy that comes with the utilization of Internet cookies generated by visiting websites or third parties.

Some websites store information in the form of plain text that can be changed whenever a user visits a web page. This information leads to the easy retrieval of cookies and can be easily forged. Cookies possess a number of threats such as network threats, end system threats, and cookie-harvesting threats. Network threats arise from the transfer of cookie data as clear text that can be altered or spoofed during the transfer. However, Secure Socket Layer can be used to protect cookies from network threats. End-system threats correspond to the impersonation of other users.

Similarly, cookie harvesting corresponds to impersonating a legitimate website and collecting cookies from users of that website. Cache sniffing is a significant security concern, and this occurs when an attacker accesses the browser or proxy cache along with the cookie content. Moreover, there is a possibility that a website or web application can gather data from a user maliciously using cross-site scripting or cookie sniffing.

Nowadays, some software companies are developing their products based on protecting a user from the threats that come with cookies. In the same way, there are products that can handle the specific type of cookies (known as tracking cookies) that are distributed, shared, and read across more than one unrelated website. The primary purpose of tracking cookies is to gather information or potentially to present customized data. These cookies are not malware, worms, or viruses but are the ones that disclose the data that correspond to user privacy. For example, when a user browses an online advertising website, the site can place a cookie on the user's device. If another site also has advertisements from the same vendor from the site he visited earlier, then that vendor knows that the user has visited both websites. So, the advertising or marketing company will indirectly determine all the sites the user visited.

Some other security and privacy concerns are about how cookies work and the types of cookies. The European Union (EU) directive of May 2011 on privacy and data protection requires inform users that cookies are not used to gather information unnecessarily. This regulation is applicable since December 11, 2018 [37] other legislation required that websites get consent from visitors to store or retrieve any information on a computer.

Not much research has been done to know how the impacts of the disclosure and the policy of using Internet cookies as tactics for non-voluntary identification may impact online users' conduct or behaviour.

A 2008 study [38] investigated the effects of cookie usage disclosures on consumer trust in the online environment from 2000 to 2007. The outcomes of the research reveal that both cookie usage and disclosure have increased, and adverse effects of consumer trust issues created by cookie detection can be reduced to some extent by prior disclosure from the website. Englehardt et al. [39] investigated how third-party cookies could be used to track a person's online activity. They found that it was possible to reconstruct 62%–73% of a typical user's browsing history.

3.9.3 Cookies, Privacy, and Regulations

Tirtea et al. [40] reviewed some security and privacy considerations and discussed the general workings of cookies, their types, privacy concerns related to cookies, and the security concerns of cookies from the EU directive of May 2011 [38], which stated that websites should get consent from visitors to store or retrieve any information on a computer.

In 2000, based on discussions and studies of the online environment, the FTC concluded that the use of undisclosed cookies is a violation of its fair information practice of notice. Since then, questions are being raised about the levels of disclosure and user privacy protections. Every year, the FTC receives hundreds of complaints regarding privacy breaches. In one of the cases, a company provided an opt-out option for cookies to users, but they still tracked user information using flash cookies. In another example, users alleged that a company promised limited tracking but used history sniffing to track the user across the web including when they visited sensitive financial and health sites. To handle such breaches, the FTC discusses their privacy policy from the perspective of social contract theory (SCT).

According to SCT, "morality consists in the set of rules governing behavior, that rational people would accept, on the condition that others accept them as well" [41]. However, the breach of the social contract using undisclosed cookies lowers the users' trust in the organization or the culpable party. The FTC issued the guidelines regarding the information gathering and usage characteristics of online privacy in accordance with user choice, that is, a user will decide how and what information he/she would like to share in agreement with the social contract. However, a user cannot choose information provisions if he/she is deprived of an organization's information-gathering practices.

3.9.4 Survey Questions

To determine the awareness of college and graduate students about Internet cookies, a survey was administered among college students who have a fundamental knowledge of Internet cookies. The survey also included questions that examined how much users are prone to security attacks due to a lack of knowledge of Internet cookies.

So, keeping these issues in mind, respondents were asked about their comprehension of Internet cookies. We assumed that the more users understand Internet cookies, the better they would be able to assess various regulatory options.

Questions
- Do you know what cookies on the Internet are and what they do?
- Do you accept cookies when browsing the Internet?

3.9.5 Results and Explanation

Thirty-two per cent (29) respondents say they know what cookies are on the Internet and understand fully how they work, 43% (39) say they have some understanding of how they work, and 17% (15) say they have heard about cookies but do not understand how they work. Finally, 8% don't know anything about cookies.

Nineteen per cent (17) of respondents accept all cookies, 54% (48) accept selected cookies, 15% (13) do not accept any cookies, and 12% (11) don't know about cookies.

This study suggests that respondents have an idea about Internet cookies but lack complete knowledge about them and the security threats they might pose. So, it is entirely possible to steal a user's information with the help of Internet cookies. That said, the responses suggest that users are cautious in accepting cookies from untrusted websites.

3.10 SOCIAL MEDIA

In today's world, the use of social networking is growing. Websites such as Facebook help people to connect with their long-lost friends, colleagues, and friends. There are various other websites for social networking serving their users. This usage is not only limited to young people or teenagers but also has spread to people of all ages. People use social networking to interact with others or to find others with similar interests. People post pictures and videos and share personal information, yet sharing is one of the primary reasons for privacy leaks. Since people reveal their desires and intentions through their posts, self-disclosure becomes a significant issue.

Many people have started giving more importance to their social/online profile than their real life. The number of likes on a picture/post sometimes is more important than an actual compliment or conversation. As people are sharing more and more of their personal information on the Internet, self-disclosure becomes a very concerning issue, as does privacy.

Social networking sites such as Facebook, Instagram, LinkedIn, and YouTube are prevalent as indicated by the numbers of users and hits per day to their sites. Specifically, for students, these sites provide opportunities to connect with people of similar interests, build and maintain relationships with friends, and feel more connected with their campus. The college environment places new academic, psychological, and social demands on students [42,43], for which the resources of local social capital would be helpful. To meet these demands, students need to rebuild or at least reshape their social networks. As is evident from their name, a primary role of social networking sites is to help individuals seek, establish, and then maintain social relationships—to build, rebuild, and regulate the sorts of social networks that provide social capital [44].

In social networking sites, self-disclosure is an essential element of this process [45]. Uncovering information about oneself on such sites as Facebook (or MySpace previously) is the reason for luring others to request to be a friend or to react to one's demand to be their friend. Once the "friend" connection has been established, self-disclosure is the basis of virtual interactions that serve to deepen the relationship [46]. Self-exposure is a procedure by which individuals share their sentiments, considerations, encounters, and information with others [47]. Social networking sites encourage disclosure of necessary information about the user, along with inner thoughts and emotional states [48]. Such disclosure is a crucial element in the process of relationship development [49]. College students tend to admire Facebook profiles with broader rather than limited information [44,46,50], although there is some expectation that more intimate information should be shared privately rather than be open to all Facebook friends.

In their study, Turban et al. [51] depicted a point of view in enterprise social networking and the clear focal points and difficulties. Five risk categories were identified and included: legal risks, security and privacy risks, intellectual property and copyrights risks, employee risks, and other risks. This report shows the actual data from the participants regarding the awareness of the risks involved in social networking sites.

Facebook is the most popular social media platform globally. According to www. Statista.com, as of the first quarter of 2018, Facebook had 2.19 billion monthly active users [52]. With such tremendous popularity come great security risks. Facebook removes 1 million accounts every day in its desperate attempt to control spam, hate speech, and fraud on the website. Despite such measures, the site remains a top target for hackers all over the globe [53].

A study reported by the Daily Mail [54] showed that Facebook logins sell for just $5.20. Facebook goes out of its way to protect user privacy by buying the passwords on the dark web and cross-referencing with their databases and alerting users to modify their password if a match is found. However, the organization has also been criticized because this move could encourage cyber-criminals [55].

Cybersecurity is also a shared responsibility; users must take precautionary measures to reduce their chances of being hacked. It has been found that half of Facebook users accept friend requests from unknown profiles. Also, users who are lax in their security measures are more vulnerable to security attacks [56].

Gross and Acquisti [57] conducted one of the first surveys into user behaviour on social networking platforms such as Facebook and Friendster by surveying 4,000 students of Carnegie Mellon University and found that 90.8% of users uploaded their images, 87.8% revealed their date of birth, 39.9% shared their phone numbers, and 50.8% revealed their addresses. In the research, they concluded that several of these behaviours could be traced to user negligence and them thinking that the benefits of giving away their information are higher than the cost. Also, the authors concluded that factors such as peer pressure, herding behaviour, and acceptance of this phenomenon as normal behaviour could be the causes for this.

Gao et al. [58] focused on four categories of security vulnerabilities: viral marketing, network structural attacks, malware attacks, and privacy breaches. The authors suggested that third-party developers request user access to personal information such as date of birth, and the users are not aware of what information is genuinely required by the application, which leads to security breaches.

Wang et al. [59] focused on how social media technologies can be used to fabricate and diffuse misinformation, and how bots operate to influence the behaviour of users. They suggested the use of a variety of techniques to differentiate between human users and bots and cross-verifying information using services such as ClaimBuster (end-to-end fact-checking system).

Approximately two-thirds of U.S. adults (68%) use Facebook, according to a survey conducted in January 2018. While a substantial share of Americans get news from Facebook and other social media sites, very few people express much trust in information on these sites. Only 5% of Americans say they have "a lot" of trust in the information they get from social media sites, while another 33% say they have "some" trust in it, according to a March 2017 survey.

3.11 FACEBOOK

Facebook is a robust connection tool. According to Statista.com, Facebook has more than 1.94 billion global monthly active users, including over close to 1.74 billion mobile monthly active users as of the first quarter of 2017 [60]. However, students do not believe in using the website for academic purposes. It seems that the low usage of Facebook for academic usage is because of the lack of proper resources that are accessible through Facebook. Also, students are looking for friends, not to find people who will help in their academics via Facebook, which means that Facebook fails as a platform for building specific academic relationships.

Since students do not tend to spend a significant amount of time on Facebook, we might conclude that students are not using Facebook for academic purposes. Unless Facebook helps them in creating connections or finding resources, they are not interested in spending more time logged in. There is a consensus among students not to share personal and private information on Facebook. Most students do not feel that the measures that Facebook takes to protect their data are enough and this leads students to refrain from posting publicly and to restrict circulating their information to only within their trusted set of "online friends."

Despite being the largest social networking platform, the lack of privacy and security has led students not to use Facebook in a way that would ultimately benefit them in academics or overall personality building.

The share of online Americans who have at least some trust in the information they get from social media (37%) is far lower than the percentage of U.S. adults who have at least some trust in the information they get from local news organizations (85%), friends and family (76%), and national news organizations (72%) [61].

According to one study, the primary needs to use Facebook are (1) the need to belong and (2) the need for self-presentation [62].

There are some reports that indicate that Facebook's user base has become older and there is more growth among older people. The younger generations and particularly college students tend to use Instagram more often [63]. This study also suggests that college students are less attracted to Facebook.

Having said that there is still a considerable population who use Facebook to share stories, pictures, get news, and more.

The 2016 U.S. election revealed the suspicions of how social media, and mainly Facebook, was used to influence the election. Much of this was based on the revelation that Cambridge Analytica, a political data firm, gained access to the private information of more than 50 million Facebook users. The firm offered tools that could identify the personalities of American voters and influence their behaviour. The case of Cambridge Analytica opened the eyes of millions of users about their privacy when using social networking tools [64]. Following this case, the media reported on many movements of people deleting their accounts or leaving Facebook.

Privacy is also a major concern, as the results of a survey have shown that 36% of the total content remains shared even if all default settings are used [65]. The study also revealed that Facebook met only 37% of the real expectations concerning their privacy settings and that almost always there was more data exposed than what was expected to on the site.

To understand how college students feel following this scandal, a group of students were asked to download their Facebook data and look at their communication on Facebook. The main aim of this study was to make them conscious of their Facebook data.

When looking at the options given to users for sharing data on Facebook, there are five different granularities: Only Me, Specific People, Friends of Friends, Friends only, and Everyone. But the recommended setting which a majority of users don't know is "Share with everyone," which means that if they do not modify their default settings, then their data could potentially be exposed to Facebook users at any given time. In addition to this, the survey suggested that when the default settings for privacy are modified according to users, only 39% met the actual requirements of the users. Thus, a great deal of information about how Facebook handles its privacy settings is still unknown to most users.

Another survey showed that only 30% of pairs (any two people at a given point in time) on Facebook interact from 1 month to the next. The survey showed that 54% of the interactions that take place between infrequently interacting pairs are only due to the birthday reminder feature provided by the site. It was also observed that 70% of the links in the activity network would disappear within 1 month. Thus, the evolution of Facebook users with the passage of time can be analysed, and studies show that there is a decline in the overall network activity, even if the number of friends on Facebook might be increasing rapidly [66].

When it comes to concerns about privacy among the youth when compared to older adults, 71% of the youth took a more considerable amount of effort to adjust their privacy settings to keep them safe on Facebook. Additionally, 20% of users in the survey said that at some point in time, they had deactivated their accounts due to privacy concerns and 82% of the youth changed their Facebook settings to make themselves less visible on Facebook in the span of 1 year. This shows that there is an increasing concern about privacy of social media. When we look at the total survey respondents, we see that about 94.9% used their real names on Facebook and only 3.5% have numerous Facebook profiles [67].

In 2005, Charlie Rosenberg, a student at the University of Missouri, wrote a computer program that enabled him to send 25,000 invites to random people to ask them to be his friends on Facebook, and that 30% of the people accepted his request without any hesitation. This showed their lack of knowledge when it comes to whether a profile is fake or authentic. A similar attempt was made by an IT security company, called "SOPHOS," where they set up a fake profile to determine if data mining was possible. To their knowledge, 41% revealed their personal information immediately, and this information was enough to create phishing emails and other harmful worms, which could produce severe damage to those unaware individuals. These surveys highlighted the fact that average users have no idea of how their data is exploited by third-party applications on Facebook. Moreover, this activity remains virtually invisible to users [68]. All of this data is fed to third parties or trickles down from the interactions and descriptions made by unaware users. The consequences are that these actions make users a greater target for identity theft. Also, such data can be reorganized and used for marketing, advertising, and public relations (PR) without the knowledge of the user.

After analysing the data from a number of worldwide surveys, research work, and papers involving Facebook statistical data, we can infer that simply reducing the visibility of users' profiles or data does not provide adequate security. There are still a number of agencies and hackers that can cause harm in the form of hacking, harassment, blackmail, phishing, and data mining. Though the gratifications of using Facebook tend to outweigh the perceived threats, the most common strategy of privacy (which is "Decreasing visibility to people") is a very weak mechanism when it comes to privacy settings. Thus, a higher level of education in computer literacy needs to be given to most adults to help them know the levels of privacy and how privacy settings have to be used to keep users safe at all times. Finally, instead of just reducing the visibility to others via privacy settings, users should systematically consider the quality of the data they share, and they should be careful when uploading content on social media sites like Facebook.

The following questions were asked about the behaviour and usage of social networking applications and Facebook in this part of the study. Some questions were asked to better understand students' behaviour in using social networking applications and some for understanding security practice.

3.11.1 Students' Behaviour

- How much time do you spend on social networking (e.g., Facebook, LinkedIn) each day?
- How often are you commenting on known people's wall/posts?
- How often are you commenting on public posts?
- How often do you respond to posts on social media?
- How important is your social media profile to you?
- How many connections/friends on social networking sites are your friends offline too?
- During the past 12 months, how many times did you share pictures, videos, or other media?

3.11.2 Security Practices

- Do you share your personal experience on social networking channels?
- How secure is your information on social media?
- What are the main criteria for you to react (like/comment) on a post?
- How often do you think about how your privacy would be affected before posting pictures, videos, etc. on social media?
- Before posting something on social media, what is your primary intention/preference?

3.11.3 Facebook Sage

To better understand Facebook usage, I asked a group of students to download their Facebook data. I told them to

- Download the data. Just go to the Settings section of Facebook (the arrow next to the question mark, top right), and click "Download my data" at the bottom of the page called "General Account Settings." Then, you'll make your request and wait for Facebook's compiling of your data, which is delivered via an email link.
- Open the Index.html.
- Create a report and time frame: from date to date.
- Make note of the number of
  - Contacts
  - Videos shared
  - Pictures shared
  - Friends
  - Send friend requests
  - Friend requests received
  - Declined friend requests

- Removed friends
- Friend peer groups.

Forty students reported the numbers and commented on their results and experiences.

3.12 SOCIAL NETWORKING— RESULTS AND EXPLANATION

3.12.1 Students' Behaviour

Thirty per cent (25) spend less than 1 h on social networking (e.g., Facebook, LinkedIn) each day, 47% (39) spend 1–3 h, 14% (12) spend more than 3 h, and 8% (7) declare they are always online (Table 3.1).

Two per cent (2) of respondents comment on known people's wall/posts all the time, 10% comment frequently, 27% (24) comment sometimes, 43% (38) comment rarely, and 17% (15) never comment or like.

Thirty-one per cent (26) of respondents never comment on public posts, 57% (47) rarely comment, 7% (6) sometimes comment, and 5% frequently comment.

Six per cent (5) of participants declare they respond to posts on social media every time they log in, 12% (10) quite often, 34% (28) sometimes, 37% (31) rarely, and 11% (9) never.

TABLE 3.1   Surveyed students' patterns of behaviour when using social networking sites

ACTIONS	SECURE BEHAVIOUR
Time spent on social networking (e.g., Facebook, LinkedIn) each day?	30% spend less than 1 h on social networking (e.g., Facebook, LinkedIn) each day 47% 1–3 h 14% more than 3 h 8% declare they are always online
Commenting on known people's wall/posts	2% of respondents comment on known people's wall/posts all the time 10% frequently 27% (24) sometimes 43% (38) rarely 17% never commenting and liking
Commenting on public posts	31% of respondents never comment on public posts 57% rarely 7% sometimes 5% frequently
Share your personal experience on social networking channels	29% never share personal experience on social media 41% rarely 22% sometimes 7% frequently

(Continued)

TABLE 3.1 (*Continued*) Surveyed students' patterns of behaviour when using social networking sites

ACTIONS	SECURE BEHAVIOUR
Security of information on social media?	19% of respondents think the information on social media is not secure at all
Respond to post on social media	6% of participants declare they respond to post on social media every time they log in 12% quite often 34% sometimes 37% rarely 11% (9) never
Main criteria to react (like/comment) on a post?	35% react (like/comment) on a post when they like the post 33% when the posted is from a close friend 17% what they raise awareness on an issue 8% when they are for political support 7% for other reasons
Think about how privacy would be affected before posting pictures, videos, and so on on social media?	24% of respondents every time they post on social media think about how your privacy would be affected before posting pictures and videos on social media 29% almost every time 27% sometimes 14% rarely 6% never
Important social media profile	15% (12) of respondents think their social media profile extremely important 10% (8) important 48% (39) important but can live without it 19% (15) not important at all 9% (7) not at all important
Primary intention/preference before posting something on social media	17% to get more like 54% portraying themselves 29% other reasons
Connections/friends on social networking sites are friends offline too	33% (27) participants declare that less than 50 connections/friends on social networking sites are their friend's offline too 27% (22) between 50 and 100 41% over 100
Number of times shared pictures, videos, or other media	4% (3) of the participant shared over 100 times pictures, videos, or other media for the past 12 months 20% (17) 50–100 times 16% (13) 20–50 times 52% (43) less than 20 times 8% (7) never shared

Thirty-five per cent (53) react (like/comment) on a post when they like the post, 33% (50) when the post is from a close friend, 17% (25) when the post raises awareness of an issue, 8% (12) when they are for political support, and 7% (11) for other reasons.

Fifteen per cent (12) of respondents think their social media profiles are extremely important, 10% (8) important, 48% (39) important but can live without it, 19% (15) not important, and 9% (7) not at all important.

Thirty-three per cent (27) of participants say they have less than 50 connections/friends on social networking.

3.12.2 Facebook Usage

Four per cent (3) of the participants shared pictures, videos, or other media over 100 times for the past 12 months, 20% (17) shared 50–100 times, 16% (13) shared 20–50 times, 52% (43) shared less than 20 times, and finally 8% (7) never shared. Respondents also provided some narrative explanation for their Facebook usage.

Students' comments on usage:

Most of the students said they found some things that they did not expect when they downloaded their Facebook data. Below is a summary of their comments (see Table 3.2 for details):

- "...Facebook holds onto the data even after app deletion..."
- "...I expected to see an aggregation of all the messages I sent according to the timestamp. This was not true, it sent me to messenger.com. The rest of the data in this section was a mess of redirects or useless data that did not reflect my activity."
- "... I could see that the messages I deleted around 7 days ago are still in the Data Archive..."
- "...Facebook was keeping track of my calls which I made with my Android phone with phone number and durations..."
- "...It was keeping track of the users I blocked which it should have just deleted..."
- "...It had at least 130 images from my previous phone, which got automatically synced to Facebook..."
- "...It kept all of my messages since the beginning, so basically Facebook had access to all of my private conversations."
- "...Facebook is not only storing my photos which I have shared, but also the IP addresses from which each photo was uploaded..."
- "...On downloading the data, I could see all sorts of data from the websites I visited but there was more to it. I saw how they are keeping track of all my activities..."
- "...Facebook keeps track of all the friend requests I've declined, removed, received, and added...."
- "...I realized that there is a lot of data that I wasn't aware of that Facebook saved."
- "...I was surprised to see that Facebook had tracked all the applications I had ever downloaded even though I had deleted the applications long ago..."
- "...I was also surprised that all the personal information of my friends/contacts was saved with Facebook and it was easily accessible to me..."
- "...Facebook not only had my current contacts (287) but also had those which I had deleted long ago."

TABLE 3.2   Students' comments on Facebook usage

Student 1

"…. Facebook recorded my activity in a marketplace and my payment history. To no surprise, I had no activity on either end…. This section of Facebook data collected information about apps I installed or used in the past. According to Facebook, I only have 5 active apps that use my Facebook data. However, I know that 2 out of these five apps are not being used anymore. It is strange that Facebook holds onto this data even after app user deletion. Another aspect of this is that I have over 60 apps that use 'expired' Facebook data…. Under Messages, I expected to see an aggregation of all the messages I sent according to the timestamp. This was not true, it sent me to messenger. com. The rest of the data in this section was a mess of redirects or useless data that did not reflect my activity."

Student 2

"… While reading my old chat history, I could see that the messages I deleted around 7 days ago are still in the Data Archive. This is very frustrating … Facebook still is able to access them… seems to be not right. The reason what Mark Zuckerberg stated while his deposition with the Senate was the data is replicated into many servers, and it takes time to delete the complete data, but he did not provide any answers related to the timeline."

Student 3

"The observations that I made were the following:

1. Facebook was keeping track of my calls which I made with my Android phone with phone number and durations.
2. It was keeping track of the users I blocked which it should have just deleted.
3. It had at least 130 images from my previous phone, which got automatically synced to the Facebook.

It kept all of my messages since the beginning, so basically Facebook had access to all my private conversations."

"… Interesting thing is Facebook is not only storing my photos which I have shared, but also the IP address from which each photo was uploaded. At the end, there is a link called 'Facial Recognition Data'. It has a field called 'Example count', which has the number of your photos Facebook has used to train their machine learning algorithm model to do facial recognition. In my case, Facebook has used 119 pictures of mine, to train their model. The "Videos" tab contains three videos, which I shared to Facebook along with the IP address and comment on each photo. Then comes the "Friends" tab, which lists down all my friends along with the month and year on which we became friends. This section also has lists of friend requests sent, friend requests received, and friend requests declined by me and list of friends I removed. I have one friend peer group i.e. Starting Adult Life.

…… The 'Ads' section lists the ad categories associated with my account. This section Applications' tab has the list of applications I have used to sign up using my Facebook account…"

(Continued)

TABLE 3.2 (*Continued*)   Students' comments on Facebook usage

Student 5

"… On downloading the data, I could see all sorts of data which I can see from the websites but there are more to it. I saw how they are keeping track of all my activities. The details like Video Shared, Pictures Shared, Friend, Sent Friend Request, Friend Request Received, Declined Friend Request, Removed Friends and Friend Peer Group are understandable or necessary for Facebook to keep as all these are part of the profile or basic details. But details like Contacts and Advertisers who uploaded a contact list with your info seems dubious to……."

Student 6

"I downloaded my Facebook….In my data…. I found it amazing how Facebook keeps track of all the friend requests I've declined, removed, received, and added. While using Facebook, I did realize that they used this information….."

Student 7

"…. I downloaded all the information that Facebook possesses with respect to me, and I realized that there is a lot of data that I wasn't aware of that Facebook saved. Also, I was surprised to see that Facebook had tracked all the applications I had ever downloaded even though I had deleted the applications long ago.
I was also surprised that all the personal information of my friends/contacts was saved with Facebook and it was easily accessible to me."

Student 8

"After downloading the HTML file of my own Facebook data, there were a number of surprising facts which I was unaware of. To begin with, Facebook not only had my current contacts (287) but also had those which I had deleted long back. All the other information like my pictures shared, videos shared, and friends were well compiled and had in alphabetical order. Another astonishing fact found in my personal information was that they had all my preferences documented and all the ads I saw were related to what I liked and how I pursued my daily life. After analyzing the type of data feed I watched, facebook also determined that my political views be 'Moderate' which came as another surprise. Also when I went through my searches on Facebook, it had all my searches according to a timeline including my deleted searches. My overall analysis after looking at the HTML file was that Facebook has way more information about me than I had ever imagined and the reason I am surprised is because I am not an avid Facebook user and rarely operate it."

Student 9

"… It has a humongous amount of information about my personal life and activities, which even now scares me as it is too personal to be shared. The friend peer group attribute entity makes you look like an object which is undergoing some experiment. It explains how basic segregation of users can be potentially utilized for targeted advertising and business analysis. For example, if it's a young girl, it will show ads of products that she could use. Similarly, this type of classification or segregation can potentially be used using for targeted advertisements, mass appeals during the elections to influence the voters. It is very difficult to get complete privacy online with the current scenario…."

(*Continued*)

TABLE 3.2 (*Continued*)   Students' comments on Facebook usage

Student 10

"… I have realized how much information Facebook has about me. A lot of information about yourself that you can't even fathom is stored. The friend peer group attribute makes you seem like a subject in some experiment. It explains how basic classification of users can be used for targeted advertising…."

Student 11

"… When I analyzed the data more closely, I realized Facebook through its messenger app on my phone had accessed my phone contact information and created a backup on my contact information in its server. I didn't realize that allowing my phone app to access my contacts allows them to copy that data. It was also overwhelming to see how everything I ever did on Facebook from 2009 was so conspicuously logged, right from the browsers that I used to log in or be the IP address that I used for every comment, post or the searches I ever made. It was interesting to see how I am being profiled and labeled on Facebook for the ads it shows or be the feed. Friend Peer Group being Starting Adult Life and 103 Advertiser uploaded on the contact list and with some suspicious names on the list were eye-opening. I also realized of giving access to many installed applications, 41 on Facebook, that I no longer use but them having access to all my data now scares me…"

Student 12

"…The most startling fact uncovered in this report was that Number of Advertiser uploaded a contact list has a massive 131 entries…"

Student 13

"… I was shocked to see that it collects almost all of my data. It's also surprising to know that Facebook is not deleting any data even if it is no longer of use to the user. It has also collected expired applications…."

When looking at these students' comments in response to finding out what Facebook has collected from them, it's interesting to note how surprised they are by their finding. It is also indicative of what people think social networking knows about them versus what social networking sites in reality collect from them. It would be important for students to become more aware of their privacy and to not fall victim to the online manipulations of opinions.

- "…Another astonishing fact found in my personal information was that they had all my preferences documented and all the ads I saw were related to what I liked and how I pursued my daily life…"
- "…Facebook also determined that my political views were 'Moderate' which came as another surprise…"
- "…When I went through my searches on Facebook, it had all my searches according to a timeline including my deleted searches…"
- "…After looking at the HTML file, it was clear that Facebook has way more information about me than I had ever imagined…"
- "…It has a humongous amount of information about my personal life and activities, which even now scares me as it is too personal to be shared…"

- "...I have realized how much information Facebook has about me. A lot of information about yourself that you can't even fathom is stored..."
- "...The friend peer group attribute makes you seem like a subject in some experiment. It explains how basic classification of users can be used for targeted advertising..."
- "...I realized Facebook through its messenger app on my phone had accessed my phone contact information and created a backup of my contact information in its server..."
- "I didn't realize that allowing my phone app to access my contacts allows them to copy that data..."
- "...Overwhelming to see how everything I ever did on Facebook from 2009 was so conspicuously logged, right from the browsers that I used to log in to or the IP address that I used for every comment, post or the searches I ever made."
- "...I am being profiled and labeled on Facebook for the ads it shows or the feed. Friend Peer Group being Starting Adult Life and 103 Advertiser uploaded on the contact list and some suspicious names on the list were eye-opening."
- "...The most startling fact uncovered in this report was that the Number of Advertisers who uploaded a contact list had a massive 131 entries..."
- "...I was shocked to see that it collects almost all of my data. It's also surprising to know that Facebook is not deleting any data even if it is no longer of use to the user. It has also collected expired applications..."
- "...I was surprised to see that Facebook had tracked all the applications I had ever downloaded even though I had deleted the applications long ago..."
- "...I was also surprised that all the personal information of my friends/contacts was saved with Facebook and it was easily accessible to me..."

When looking at these students' comments in response to finding out what Facebook has collected from them, it's interesting to note how surprised they are by their findings. It is also indicative of what people think social networking knows about them versus what social networking sites in reality collect from them. It would be important for students to become more aware of their privacy and to not fall victim to the online manipulations of opinions.

3.12.3 Security Practices

Seventeen per cent of respondents' intention when posting something on social media is to have more "likes." However, 54% (44) say their intention is portraying themselves and 29% (24) for other reasons (Table 3.3).

Twenty-nine per cent (24) of respondents declare never sharing their personal experiences on social networking channels, 42% (34) rarely, 22% (18) sometimes, and 7% (6) frequently.

Nineteen per cent (16) of respondents think that information on social media is not secure at all, 64% (53) think it's relatively secure, and 17% (14) think that information on social media is very secure.

TABLE 3.3 Students' secured and unsecured behaviour

ACTIONS	SECURED BEHAVIOUR	UNSECURED BEHAVIOUR
Participate in online surveys, promotions in hopes of receiving goodies, gift cards, or prizes	49% give limited personal information or do not participate at all.	50% participate all the time or sometimes in online survey promotion in hopes of receiving goodies, gift cards, or prizes
Likely permit third-party apps services online	60% give permission to access their first name and last name	14% give permission to access their contact and phone book 18% give permission to access their location 3% give permission to read their messages 4% give all the permissions
Grant all permissions to every application while installing	47% grant permissions which they deem necessary 27% grant some permission which they deem necessary to popular apps.	11% grant all permissions to every application while installing, while 16% grant all permissions to popular apps
The personal information willing to submit to applications/services that ask such information	43% share their first name, last name, and email only 8% do not give any information at all	11% feels comfortable to submit to applications/services their first name, last name, email, address, phone number, and other information like employment 16% just their first name, last name, email, address, and phone number 22% their first name, last name, email, and phone number only
Google location service and/or other location services active on your phone (GPS enabled)	70% sometimes on, while using maps or ride-sharing services like Uber.	23% have their Google location service and/or other location services active on their phone (GPS enabled) all the time 7% almost keep location service always on.

(Continued)

TABLE 3.3 (Continued)   Students' secured and unsecured behaviour

ACTIONS	SECURED BEHAVIOUR	UNSECURED BEHAVIOUR
Respondents on social media platforms like Facebook give permissions to entertainment-based apps like "Which Game of Thrones character you are"	82% don't give permissions	18% on social media platforms like Facebook give permissions to entertainment-based apps like "Which Game of Thrones character you are" since they think it is fun.
Likely to accept app/game invites on Facebook	77% never accept a game invitation	4% generally accept app/game invites on Facebook to check out what's the deal 19% only sometimes accept app/game invition
Give permission to use the location to apps on the phone	31% only to the app that they trust 48% only for necessary services like ride sharing 11% don't give permission at all	10% participants give all the app permissions to use their location on their phone
Privacy settings on a browser	25% leave the default settings on their favourite browser without checking 10% review the default settings but do not change them 42% change the privacy settings to give themselves more privacy 7% change settings for less privacy 17% don't remember what they did	25% leave the default settings on their favourite browser without checking 10% review the default settings, but do not change them 42% change the privacy settings to give themselves more privacy 7% change settings for less privacy 17% don't remember what they did
Know what are cookies on the Internet and what they do	32% of respondents declare they know what cookies on the Internet are and understand fully how they work 43% of respondents have some understanding of whom they work	17% have heard about cookies but did not understand how they work 8% don't know anything about cookies

(Continued)

TABLE 3.3 (Continued) Students' secured and unsecured behaviour

ACTIONS	SECURED BEHAVIOUR	UNSECURED BEHAVIOUR
Accept cookies when browsing on the Internet	54% accept selected cookies 15% do not accept any cookies	19% of respondents accept all cookies 12% don't know about cookies
Aware of 2FA/two-step authentication and its benefits	83% of respondents declare that they are aware of 2FA/two-step authentication and its benefits	14% are not aware 3% are not sure
Use of 2FA on your online accounts	25% of respondents use 2FA on all their accounts when available 34% use 2FA on some of their accounts when available	19% (23) use only when it is required 12% do not use 2FA 10% declare they do not know what it is
Often give social security number while applying for job	36% never share their social security number	25% always share social security number while applying for job 20% share most of the time 3% share half of the time 17% share sometimes
Often check financial statements to ensure there is no suspicious activity before credit card bill payment	19% of respondents regularly check their financial statements to ensure that there is no suspicious activity before credit card bill payment 30% most of the time 35% always	10% never review their statement. 28% most of the time 9% half of the time
Often do you log out of your account from shared devices?	65% declare they always log out of their account from shared devices	14% about half of the time 11% most of the time 10% never

(Continued)

TABLE 3.3 (Continued)   Students' secured and unsecured behaviour

ACTIONS	SECURED BEHAVIOUR	UNSECURED BEHAVIOUR
The cost willing to pay if computer got infected by ransomware	54% (57) of respondents say they won't pay any money if their computer got infected by ransomware tomorrow	24% accept to pay if the cost were under $100 13% pay between $100 and $500 6% pay between $500 and $600 3% ready to pay over $1,000
Taking regular backups to avoid paying ransom if your computer is affected by ransomware	36% (38) of respondents take regular backups to avoid paying ransom if their computer is affected by ransomware	33% don't make any backup 31% (33) back up once in a while
Password reuse	23% do not reuse a password	39% reuse a password 39% sometimes
Often do you change your password?	7% change password after over 6 months 29% every 3–6 months 13% less than 3 months	51% of respondents change password only when is required
Share passwords	80% of respondents do not declare they do not share passwords	15% sometimes 7% say they do share passwords
Often update application on your cell phone	31% update application on your cell phone always, immediately when available 28% use automatic updates	34% once in a while 8% rarely
Grant all permissions to every application while installing	47% grant permissions which they deem necessary 27% grant some permissions which they deem necessary to popular apps	11% grant all permissions to every application while installing 16% grant all permissions to popular apps

(Continued)

TABLE 3.3 (Continued) Students' secured and unsecured behaviour

ACTIONS	SECURED BEHAVIOUR	UNSECURED BEHAVIOUR
Personal information willing to submit to applications/services that ask such information	8% do not give any information at all 43% first name, last name, and email only	11% (10) of respondents feel comfortable to submit to applications/services their first name, last name, email, address, phone number, and other information like employment 16% just their first name, last name, email, address, and phone number 22% their first name, last name, email, and phone number
Google location service and/or other location services active on your phone (GPS enabled)	70% sometimes on, while using maps or ride-sharing services like Uber	23% have their Google location service and/or other location services active on their phone (GPS enabled) all the time 7% almost keep location service on
Give permission to use your location to apps on your phone	31% only to the app that they trust 48% only for necessary services like ride sharing 11% don't give permission at all	10% participants give all the app permissions to use their location on their phone
Know what are cookies on the Internet and what they do	32% know what cookies on the Internet are and understand fully how they work 43% have some understanding of whom they work	17% had heard about cookies but did not understand whom they work 8% don't know anything about cookies
Do you accept cookies when browsing on the Internet?	54% accept selected cookies 15% do not accept any cookies	19% of respondents accept all cookies 12% don't know about cookies

(Continued)

TABLE 3.3 (Continued) Students' secured and unsecured behaviour

ACTIONS	SECURED BEHAVIOUR	UNSECURED BEHAVIOUR
Security of information on social media?	19% of respondents think the information on social media is not secure at all	64% relatively secure 17% think the information on social media is very secure
Think about how privacy would be affected before posting pictures, videos, etc. on social media	24% of respondents every time they post on social media think about how your privacy would be affected before posting pictures, videos, etc. on social media 29% almost every time 27% sometimes 14% rarely 6% never	14% rarely 6% never

Twenty-four per cent (28) of respondents think about how their privacy would be affected before posting pictures, videos, etc. on social media every time they post, 29% almost every time, 27% sometimes, 13% rarely, and 6% never.

In a time where social networking is at its peak, it is highly essential that the Internet remains a safe place for people of all age groups. This way, everyone can happily enjoy the perks of social networking without facing the stress of cyberbullying, information leaks, or any such activity. This study suggests that although people are aware of the underlying facts, they tend to take the easy way out. People need to be made more aware of the risks that come with the perks of social networking.

An overall glance of the above data indicates how much data is collected by social networking applications and the lack of awareness in even those individuals who consider themselves aware of cybersecurity. It seems that generally, the gratifications of using Facebook tend to outweigh the perceived threats and dangers.

3.13 CONCLUSION

Table 3.1 summarizes the results from the social networking survey and the students' patterns of behaviour when using social networking sites. It appears that almost half of respondents spend 1–3 h on social networking, 43% rarely comment on known people's wall/posts, and 27% sometimes comment. The main reason people react to a post is when they like it (35%) or when the post is from a close friend (33%). Respondents rarely or sometimes respond to posts on social media (71%). A quarter of respondents think about how their privacy would be affected before posting pictures or videos on social media. A quarter of respondents think their social media profile is extremely important or important to them. The primary intention/preference of over half of the respondents before posting something on social media is portraying themselves. Over half of the respondents (52%) shared less than 20 times for the past 12 months.

Overall, it seems that college students are interacting less in social media. In the survey, we did not differentiate the different types of social media (Facebook, Snapchat, Instagram, and so on); thus, the numbers represent usage of all types of social networking. However, it appears while some students heavily use social media, others are using it less or at least declaring to use it less.

Faculty Cybersecurity Awareness

4

4.1 INTRODUCTION

After conducting the general survey with students, the following question emerged: What would be the results if we had asked the same questions of university faculty? Would the results be very different? Since most universities in the United States require the faculty to take online courses on cybersecurity, it was assumed that the knowledge of cybersecurity and the actions they take to protect themselves would be better than the students. Thus, the same general survey was sent to the faculty of a public university in the Bay Area of California. Overall, 63 faculty members completed the survey. Twenty-one per cent aged 24–34, 23% aged 35–44, 16% aged 45–54, and 40% aged over 55. Thirty per cent were assistant professor, 5% associate professor, 34% full professor, and 32% part-time/adjunct professor. Fifty-seven per cent of respondents were female and 43% male. Despite the composition of the respondents, the number of respondents was not enough to analyse the data in smaller subgroups.

4.2 KNOWLEDGE OF SECURITY

In response to the question regarding faculty's knowledge of the concept of cybersecurity, only 43% agreed that they are knowledgeable (agree or strongly agree), 44% believed they have average knowledge (somewhat agree or neither agree nor disagree), and 13% reported they are not knowledgeable (somewhat disagree or strongly disagree).

Interestingly, both faculty and students self-evaluated themselves quite close to each other, with slightly more faculty in the category of "Agree." Otherwise, they were almost the same in all other levels. Only 6% of faculty believed that they are knowledgeable about cybersecurity (see Figure 4.1).

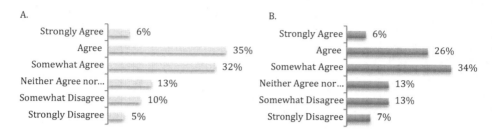

FIGURE 4.1   Chart A shows faculty knowledge on cybersecurity (63 respondents) and Chart B students' knowledge on cybersecurity (367 respondents).

4.3 PRIVACY

The faculty was asked to select the type of information that they consider to be private. The data that respondents considered most private was bank account information (95%), followed by contact information (68%), pictures (66%), online search activity (50%), internet protocol (IP) address (50%), location (47%), and others (13%). Overall, it seems that users consider bank information, contact information, and pictures as private information.

Again, the faculty members think much in the same way as students (see Table 3.2). IP address and locations that were more sensitive were ranked lower. Considering these important elements as less private might indicate the lack of awareness of the significance of the privacy of each (see Figure 4.2).

The respondents were also asked to rate the security of each platform on a scale of 1–10 (1 being the least secure and 10 being the most secure). In this case what was ranked higher (strongly agree and agree) were as follows: online banking (64%), mobile banking (47%), online shopping (31%), texting (25%), email (14%), and social media (4%).

Overall, the data suggests that

- Forty-seven per cent of university faculty agreed that online banking is secure, with only 7% who disagreed or strongly disagreed overall and 5% neutral.
- University faculty further agreed that online shopping is mostly secure (31%), 26% disagreed and 7% were neutral.
- Social media was considered least private with a considerable disagreement of 57% and low agreement of 4%. Neutral reviews were 5%.
- Twenty-five per cent of faculty respondents agreed that texting is secure. Eighteen per cent of faculty has a neutral "Neither agree nor disagree" that texting is secure. Messaging apps were similar to texting, where 15% agreed they are secure, while 30% disagreed. Neutral reviews were 20%.

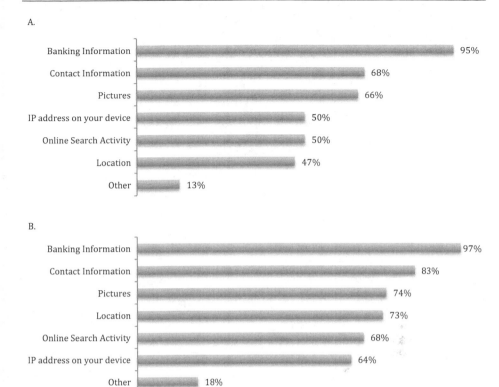

A.

B.

FIGURE 4.2 Chart A shows faculty consider private information (63 respondents) and Chart B students consider private information (367 respondents).

- Fourteen per cent of faculty respondents agreed that email is secure, and 25% disagreed. Neutral reviews were 17%.
- People were somewhat positive about security in online shopping (31% somewhat agree).

For the question of whether or not respondents had ever rejected a mobile app request for accessing their contacts, camera, or location, 54% responded: "Yes." The "No" responses were 41%, and 5% replied, "Not sure."

4.4 PASSWORD

Regarding password selection, the faculty was asked if they use a harder-to-guess password to access their bank account than to access their social networking accounts. Sixty-one per cent said "Yes" and used a harder-to-guess password for their bank account, 3% used the same type of complexity for both passwords, and 36% used the same password for both types of accounts.

Additionally, 50% of respondents said that they used two-factor authentication (2FA) for some accounts, 30% used it for all accounts, 10% didn't use it at all, and 10% didn't even know what it is. Thus, 20% of respondents didn't know or didn't use 2FA.

4.5 TRUST

An important factor in security is the perception of trust in computer systems, which was evaluated first through a basic security question about trusting the Internet: "Do you have reason to believe that you are being observed online without your consent?"

Forty-seven per cent of respondents had reason to believe that they were being observed online without their consent, 28% had no reason to believe, and 25% were not sure. These percentages might be indicators of the overall trust of Internet privacy.

4.6 TRUST OF UNIVERSITY DATA SECURITY

We also asked two other questions about trust, this time specifically regarding trust in the university system: "Do you think that your data on the university system is secure?"

Only 7% believed that their data was secure, 67% believed that data was relatively secure on university systems, 18% declared not secure, and 8% declared not sure.

The other question was, "Do you think your communication through Learning Management System is secure?" and the responses to this included 7% of faculty who think it is secure, 54% relatively secure, 12% not secure, and 28% not sure.

4.7 CONCLUSION

The survey results from the students and faculty are summarized in Table 3.3. Looking at all the factors, one can see that faculty and students are almost the same when it comes to their beliefs or actions. There are no major differences between the two groups (see Figure 4.3). Thus, this study highlights that faculty are not necessarily thinking or behaving differently than their students.

The number of faculty participated in the survey was not large enough for further subgroup analysis. A follow-up survey with faculty was not conducted as it was with students. However, we would feel confident in predicting that there would not have been a significant difference between faculty and students had a follow-up survey been conducted.

If we think of faculty as university employees who are required to take online trainings on cybersecurity, then the results of this survey are somewhat discouraging. It seems that a tremendous effort needs to be undertaken by university officials to fundamentally change this trend by providing a more systematic and informative awareness programme for employees at all levels, and particularly the faculty. This strategy would not only benefit the faculty directly but also indirectly help students and the university as a whole (Figure 4.3 and Table 4.1).

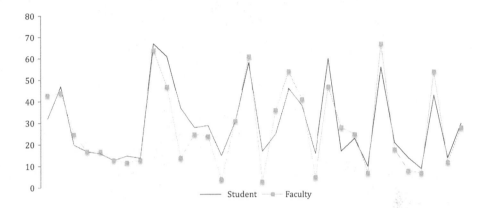

FIGURE 4.3 Comparison between surveyed students and faculty cybersecurity awareness (based on the data of Table 3.3).

TABLE 4.1 Comparison between surveyed students and faculty cybersecurity awareness

AWARENESS	STUDENTS	FACULTY
Knowledgeable about the concept of cybersecurity	32% agree or strongly agree 47% average knowledge, somewhat agree, or neither agree nor disagree	43% agree or strongly agree 44% average knowledge, somewhat agree, or neither agree nor disagree
Consider being private information	20% bank account information 17% contact information 16% pictures 13% IP address on the device 15% location 14% online search activity	25% bank account information 17% contact information 17% pictures 13% IP address on the device 12% location 13% online search activity
Security of each platform (strongly agree and agree)	67% online banking 61% mobile banking 37% email 28% texting 29% mobile texting 15% social media 31% online shopping	64% online banking 47% mobile banking 14% email 25% texting 24% mobile texting 4% social media 31% online shopping

(Continued)

TABLE 4.1 (*Continued*)   Comparison between surveyed students and faculty cybersecurity awareness

AWARENESS	STUDENTS	FACULTY
Harder-to-guess password to access your bank account than to access your social networking accounts	58% use a harder-to-guess password for their bank account than for social networking 17% use the same complexity for both passwords 25% use different passwords but the same complexity for both	61% use a harder-to-guess password for their bank account than for social networking 3% use the same complexity for both passwords 36% "No, use the same for both"
2FA usage	54% of respondents use 2FA for some accounts, and 27% use for all accounts 12% do not use it 6% "No, don't know what it is"	50% of respondents use 2FA for some accounts, and 30% use it for all accounts 10% do not use it at all 10% do not know even what it is
Mobile app permission request to access contacts, camera, or location?	46% "Yes" 38% "No" 16% "Not sure"	54% "Yes" 41% "No" 5% "Not sure"
Do you have reason to believe that you are being observed online without your consent?	60% believe that they were being observed online without their consent, 17% had no reason to believe, and 23% were not sure	47% believe that they were being observed online without their consent, 28% had no reason to believe, and 25% were not sure
Security of data on the university system	10% believed that their data are secure 56% relatively secure 21% not secure 14% not sure	7% believed that their data was secure 67% relatively secure 18% not secure 8% not sure
Security of data on the Learning Management System	9% believed secure 43% relatively secure 17% not secure 30% not sure	7% believed secure 54% relatively secure 12% not secure 28% not sure

How to Improve Awareness

5

5.1 INTRODUCTION

The results of the reported investigations in this book suggest that to improve the security of people in cyberspace, knowledge and awareness plays a vital role in enhancing security. We need to understand that simply having knowledge is not enough. As the famous Chinese proverb from Confucius puts it: "The essence of knowledge is, having it, to apply it." Consequently, the knowledge given to people must be applied so that they can benefit from it in their cyber lives. Our survey results indicate that even when people know about cyber risks, they are still not always taking practical actions to protect themselves. In this chapter, we will review the literature on awareness and training programmes and then discuss some potential solutions.

A significant number of security breaches are due to users' lack of knowledge or unsafe behaviours such as sharing passwords and opening unknown emails and attachments. These activities potentially open up the organization or individuals to threats from hackers and to the loss of assets of individuals and organizations.

Although organizations and enterprises invest and rely more on technology for security solutions (e.g., firewalls, antivirus software, and intrusion detection systems) and to defend organizational assets, the importance of considering the role of users in the security equation has grown. On the one hand, users/employees need to understand security issues, while on the other hand, they must follow the security policies of each organization, which become crucial to comply with information security (IS) laws and regulations.

Despite the role of users, there are huge investments in defensive technologies, but little investment in human awareness.

The first step in improving awareness is to measure it among each targeted group. The aim of awareness measurements is to use a reliable methodology to measure awareness in cybersecurity. Combinations of three different methods are used to measure awareness mostly among employees of a company. Questionnaires and surveys are used to measure knowledge (what you know), attitude (what you think), and behaviour (what you do) [23].

Model-driven techniques and survey-based research are also used to investigate behaviour modelling in the security context such as information-sharing and security policy compliance [69] as well as computer security behaviour while interacting with email attachments [70].

Egelman and Peer [71,72] developed the Security Behavior Intentions Scale (SeBIS), to measure the intention of security rules that end users employ while interacting with a wide variety of security controls and interfaces.

To improve the awareness of employees, many vendors have gradually formed several programmes. However, there are not many studies investigating the effectiveness of their outcomes. Most of the existing studies focus on three areas: awareness (measurements, security training, content for security training), delivery methods, and programme effectiveness measurements. In the following section, we will review each area.

5.2 TRAINING AND EDUCATION

Looking at all types of training offered, we classify them in the following four groups: formal education, professional training, employee training, and people training.

5.2.1 Formal Educational Programmes

Different formal education programmes are offered as a degree or specialized courses to train future professionals. Even though the number of educational programmes to educate information technology (IT) professionals is not enough to match the demand, they are gradually growing in numbers and continue to evolve and improve. Most cybersecurity educational programmes require at least a bachelor's degree in a related field to get started. The content of each programme mainly includes specific technical areas such as the following:

- Legal issues in information assurance
- Network programming
- Secure electronic commerce
- Discrete structures
- Computer forensics
- Audits and regulations
- Cryptology
- Computer security
- Database security
- Database design.

These types of educational programmes are not the focus of this book since their objectives are to educate and train IT professional.

5.2.2 Training Programme for IT Professionals in the Industry

A second group of educational programmes includes the hundreds of training workshops, certificate programmes and tutorials of all kinds offered by private companies, training institutions and individuals on a variety of technological tools, technical topics, and solutions. These types of educational programmes are also not the focus of this book since their objective is to educate the professional workforce.

5.2.3 Employee Training

With the growing number of companies that have become victim to cyber-attacks, the protection of information systems and information assets from cybersecurity threats has become critical. Despite this fact, most companies do not provide training but are still more reliant upon technology to resolve their cybersecurity issues. Thus, employees often lack cybersecurity knowledge and skill sets and are identified as susceptible threat vectors by cyber-attackers and are, therefore, being targeted with continually evolving threats [73].

Despite considerable investment in organizational security, the majority of the approaches and protection methods focus heavily on external attacks and technological defences and have not minimized the number of security incidents [74]. However, Abawajy [75] points out that the organization is only as secure as its weakest link. Stanton et al. [76] stated that even the best technology efforts intended to address IS would fail unless the organization's employees take the proper courses of action when approached with a threat.

Although technology-oriented safeguards such as firewalls and intrusion detection systems are found in a large number of organizations, the focus on human factors in security including awareness and training initiatives has historically lagged behind [77]. Previous studies in IS literature have confirmed awareness techniques to be effective in increasing employee security-related knowledge and promoting security-conscious decision-making. However, the benefit of an educated general business community is limitless [78]. Technology alone cannot solve a problem that is controlled by individuals.

5.2.3.1 Security Education, Training, and Awareness programmes

A Security Education, Training and Awareness (SETA) programme is designed to reduce the number of security breaches that occur through a lack of employee security awareness. These programmes mainly explain the employee's role in the area of IS. A SETA programme is generally offered as part of the employee orientation programme. The main content of a SETA is to explain each organization's security policies.

5.2.3.2 Cybersecurity countermeasures awareness

Cybersecurity countermeasures awareness (CCA) is the state where individuals are aware of their cybersecurity mission within the organization. In general, a CCA programme includes SETA programmes, computer monitoring, and various security countermeasures to make people aware of their cybersecurity mission within the organization.

5.2.3.3 Cybersecurity skill

This programme aims to improve an individual's technical knowledge, ability, and experience surrounding the hardware and software required to execute IS in protecting IT against damage, unauthorized use, modification, and/or exploitation.

5.3 SECURITY TRAINING CONTENT

While training programmes and initiatives exist within many organizations, there appears to be limited empirical research to determine which topics should be covered, what the most useful delivery methods are, and to what degree these factors play a part in the IS practice of employees [79].

A study of 252 global organizations found nine principal cyber-attack vectors, most of which focused on the human factor in IS including viruses, malware, web-based attacks, phishing and social engineering, malicious code, denial of services, and stolen devices [80].

Comparative research among the different training programmes concluded that there are no statistically significant mean differences on employees' CCA and cybersecurity skills (CyS) between the two SETA programme types (typical and socio-technical) [81].

5.4 DELIVERY METHODS

The training delivery methods to improve awareness among employees or IT professionals are the subjects of several studies. Even though there are no conclusive reports that illustrate what approach would be the best, there is some indication of what would work better and these are worthy subjects to the discussion. Overall, the following training delivery techniques can be identified:

- Face-to-face training or classroom
- Online training
- Online instructor-led training
- Games
- Competitions
- Mass media: posters, emails, podcast newsletter, etc.

Abawajy (2012) reviewed the literature and investigated the user preferences of cybersecurity awareness delivery methods. Although people often express interest in a classroom-based delivery method, this is relatively expensive and provides a "static solution for a fluid problem" [82]. Also, many users find it to be boring and ineffective [83]. In general, the success of classroom training depends upon the ability of the instructor to engage the audience. It might also tend to fail because it is based on rote memorization and does not require users to think about and apply IS concepts [84].

Sharing experiences and knowledge between the employees of an organization facilitated by participation makes classroom training more effective [85]. However, this approach assumes that participants are knowledgeable about the subjects being discussed.

Despite the extensive use of online training sessions in many companies, there is no conclusive evidence that online standardized courses are effective enough to create a successful security culture in an organization.

The use of games to train people has been growing. Games are good tools to motivate people to focus their attention on specific issues. However, no evidence has been found on how effective they are. "Serious games" (defined as games with a purpose other than pure entertainment) are used in training IT professionals to explore solutions or issues, but these are not effective in educating ordinary individuals.

5.5 TRAINING AND AWARENESS PROGRAMME EFFECTIVENESS

No matter what type of training is used to improve cybersecurity awareness, it is more important to measure its success in not only educating employees of their knowledge of cybersecurity, but also to see if and how they integrate this awareness into their everyday practice and behaviour. It is also essential to measure the effect of awareness training on the actual behaviour of the trainees.

Egelman et al. developed the SeBIS [86], a 16-item, scale-based instrument to measure the intention of security rules that end users employ while interacting with a wide variety of security controls and interfaces [87].

Some studies are conducted to measure programme effectiveness. For example, Parsons et al. [88] have developed a survey instrument (Human Aspects of Information Security Questionnaire – HAIS-Q) including a 63-item measure that assesses seven focus areas: password management, email use, Internet use, social media use, mobile devices, information handling, and incident reporting. Each focus area is further divided into three specific subareas resulting in 21 areas of interest, each of which is measured via a separate knowledge, attitude, and behaviour.

5.6 END USERS' TRAINING AND AWARENESS

In regard to end users, there have been field studies done to understand novice users' views about security practices and awareness [89]. An early field study on security-related employee behaviours, such as backup and file access practices, indicates knowledge and informal heuristics as better determinants of behaviour than enforced security policies. Such qualitative investigations (interviews and field observations) enable an in-depth exploration of a narrow work domain, context, or demographics, but results from these may not be applicable to the larger population.

There is no specific training on how to train individuals to protect themselves in cyberspace aside from what exists in mainstream press and media.

5.7 A NEW APPROACH TO AWARENESS PROGRAMMES: ISSUES AND CHALLENGES

5.7.1 Cost of Security Awareness Training Programmes

According to www.infosecurity-magazine.com [90], the cost of security education for large enterprises is $290,033 per year, and about 94% of chief information officers surveyed have pushed for increased investment in user education following the recent headlines regarding phishing and ransomware. Almost all of them (99%) see users as the last line of defence against hackers, which means that user education, policies, and procedures are essential to ensuring that employees understand their role.

From the research point of view, no reliable and systematic study has been found that indicates how much this investment improves security. No data seems to be available in estimating the cost to develop security awareness for the general population. However, according to 2018 Identity Fraud: Fraud Enters a New Era of Complexity from Javelin Strategy & Research [91], in 2017 there were 16.7 million victims of identity fraud, a record high that followed a previous record the year before. Criminals are engaging in complex identity fraud schemes that are leaving record numbers of victims in their wake. The amount stolen hit $16.8 billion in 2017 as 30% of U.S. consumers were notified of a data breach, an increase of 12% from 2016. For the first time, more social security numbers were exposed than credit card numbers.

Numerous surveys and research studies have shown that ransomware is another area of growing concern for businesses, governments, and personal threats. While there is no precise data about how much this costs individuals, it has been reported that each occurrence yields an average of $1,077 for criminals [92].

Knowing that the cost of cyber-attacks is exponentially growing for the population, if we add the costs of all identity thefts, ransomware, and all enterprise costs of data breaches due to human error, we see that these costs far surpass the amount of money being invested in raising awareness in the overall population.

5.7.2 Changing People's Behaviour

In order to improve people's awareness of cybersecurity (not just their knowledge but also how to implement better behaviours into their everyday lives), one must come up with structural solutions that address people at an early age. Much like other behaviours ensuring safety and security (such as locking their cars or homes or keeping their valuables in a safe box at home or in a bank), people should also learn to lock their online home/account effectively or safeguard their digital assets. If people use a good lock for their home or business or place cameras all over the place, why shouldn't they learn to do the same for their digital assets and to monitor their online activities?

To effectively improve awareness and change behaviour, the population needs to be educated very early, continuously, and consistently. After all, it is unlikely that issues in cybersecurity will be resolved soon. Thus, it is in the interest of the greater community to include measures of cybersecurity in the educational system from middle school and up. Because children are using computers at a very early age these days, it is that much more important that they be aware of their cybersecurity as early as possible.

Including cybersecurity in the formal education system would also be the more cost-saving approach for society considering all the costs associated with training employees and individuals following instances of identity theft, ransomware, and so on.

5.7.3 Cybersecurity 101: A Solution

To improve cybersecurity awareness and prepare students for professional life, it would be extremely effective to include a basic course on cybersecurity in all university education curricula. After all, most universities require students to take many introductory-level courses such as English or Chemistry 101, so why wouldn't we offer cybersecurity awareness in all disciplines of an education curriculum? This approach would be the most effective way to fundamentally help students be prepared to protect themselves, and the organizations that they will work for, in the future. Instructors could then prepare and update their content based on the evolving and constantly changing protection methods to keep students updated about cyberspace and how to practise safe behaviour. This approach would also help educational institutions to protect universities from cyber-attacks. The content of this type of education could include, but would not have to be limited to, the following areas:

1. Trust
2. Authentication
3. Privacy
4. Ransomware
5. Identity theft
6. Phishing
7. Application access
8. Social media
9. Social engineering
10. Surveillance
11. Mobile device protection

The content could include the awareness and practice of each topic. As a result, people could learn about each area and also the ways they could implement their knowledge.

5.7.4 Cybersecurity Games: Another Solution

With the expansion of online and video games and their massive popularity among the younger population, creating scenario-based gaming is another effective way to improve overall awareness among the population. After all, many people (especially younger people) spend a great deal of time playing video games. There are several games available for IT professionals and cybersecurity that can create more dynamic learning tools. There are also games targeting the younger population, the majority of which are available for free [93].

However, large audience computer games dealing with cybersecurity issues running on main gaming platforms are not yet available on the market. After all, if there are hundreds of wargames available on all gaming platforms that show how to pilot an airplane, why not create cybersecurity games? Studies have shown that games can not only be effective training tools but also be effective for encouraging behavioural change [94].

The author has created a few card games, board games, and even online games that are used in his classes and training programmes, and he has been pleasantly surprised by their effectiveness and the degree to which students are not only having fun but also engaging with the topic. This approach could also expand to cybersecurity awareness among employees. Many people use games like Solitaire or other sorts of card games on their computers, so why not let employees play a fun game that would let them learn cyber awareness and safe behaviour? This would surely be better than just asking them to listen to static online trainings. Couldn't this be a better approach and more cost-effective? Couldn't this be an area for private enterprises to invest in or for the National Science Foundation to promote research and development?

5.7.5 Cybersecurity Culture

The numerous cybersecurity threats will not be solved in the near future simply by technological solutions. The speed at which technologies evolve in finding solutions is not as fast as the complexity of cyber-attacks. The number of Internet-enabled devices in homes, cars, and cities makes keeping up with security and protection more challenging. The events in the last U.S. election and interferences in mass communication by what is now labelled as "fake news" make trust between what is real and fake a challenging issue for citizens. The example of Cambridge Analytica [4], a political data firm that gained access to the private information of more than 50 million Facebook users and then offered tools that could identify the personalities of American voters and influence their behaviour, is a good illustration of how much citizen privacy is at stake. It also shows the importance of the judgement of people in what they choose to share.

5.8 CONCLUSION

In conclusion, in order to elevate awareness, cybersecurity needs to become a "security culture" both at work and at an individual level. Cybersecurity is not a one-time solution, and it should not be reduced to just strong passwords or a few protective approaches. It needs to become an individual responsibility. Knowing about security is vital. By addressing security concerns and risks, students and all individuals can better protect themselves against risks at work and at home.

Conclusion

6

The results of the surveys presented in this book show that the assumption that college students and faculty have a higher level of cybersecurity knowledge and can better secure themselves is incorrect. For example, the data indicates that among people who consider themselves knowledgeable, only 52% use two-factor authentication (2FA) for some accounts and they do not have a secure password, or that a surprising 8% do not even know what 2FA is. The students and faculty who think they are being observed online still are not acutely aware of how to protect their data. They report low levels of 2FA usage or password complexity for accounts, privacy settings, or protection against social engineering.

Other indicators that illustrate the lack of sufficient awareness and protective measures include monitoring privacy policies and location access, giving permissions to apps, evaluating the trust in using applications, and the awareness of cybersecurity issues when using social networking sites.

Sixty per cent of college students in Silicon Valley, California, self-evaluate their knowledge of cybersecurity as "good" to "very good," while 25% have "no knowledge of cybersecurity" and 15% are neutral. Although the 60% seems to be a good sign, it is still very alarming to observe that around 25% do not have much knowledge of cybersecurity. If this is the case in one of the most technologically developed locations in the world, then one might assume that this percentage must be much higher among the general population or other areas. The same conclusion would likely hold true for faculty as well.

Also, it appears that educational institutions do not have an active approach to improve awareness among college students to increase their knowledge of these issues and how to protect themselves from potential cyber-attacks, such as identity theft or ransomware. It also appears that most students are aware of the possible dangerous consequences of providing personally identifiable information, such as identity theft and stalking, to a university, but they nevertheless feel comfortable enough to present it.

Security awareness is considered the first line of defence for the security of information and networks. Consequently, incorporating training to improve security awareness among college students, and even earlier at the high school level, seems to be extremely important.

The perception of trust in university data security is not very high since only 65% consider data on the university system as "secure" or "relatively secure." If this is the case in Silicon Valley, what would be the perception of students elsewhere?

To provide a general image of student awareness, I have created four personas that summarize the main user profiles in terms of cybersecurity awareness. These personas help to create user-centred training and to create awareness material such as educational poster articles or other such tools. Readers should be advised that these personas are not real people. They are just hypothetical portrayals of students in Silicon Valley, based on the information obtained in the surveys.

6.1 PERSONA 1: CHRIS

Chris is 20 years old and is an undergraduate student in his sophomore year. He studies software engineering. Chris is very knowledgeable of cybersecurity, reads the news, and is very aware of protecting himself and his personal information in cyberspace. Chris is very sensitive to privacy. In general, he uses a hard-to-guess password for his accounts, clears the cookies on his favourite browser every so often, uses 2FA when it's required, and updates the apps on his phone when available.

He does not frequently use Facebook or other social networking sites and rarely posts or shares personal information.

6.2 PERSONA 2: SAMANTHA

Samantha is a 24-year-old graduate student in psychology and is not very computer savvy. She is a fan of all social networking activity, in posting, commenting on friends' posts, liking, and so on. She is always online and has downloaded many apps on her mobile phone, which she uses and enjoys.

Samantha considers herself a lazy computer user as she never updates her apps on her computer or mobile device. She frequently forgets the passwords she creates so she has decided to use two simple passwords that she can remember.

6.3 PERSONA 3: JAMES

James is a senior undergraduate student and is about to graduate with a degree in social sciences. His intuition is to work as a researcher for a couple of years and then continue his studies at a graduate level. James is very actively involved in several political communities both on and off campus. He is constantly online chatting, sharing stories, or discussing issues. James' days are very busy, and he does not have much time to do things like improving his computer skills. He recently dealt with a phishing email attack where he clicked on a phishing email that ultimately infected his computer. He lost a lot of data and was forced to reformat his computer. Despite this incident, he still struggles to back up his computer, protect his account, update his applications, or even check his bank account statement. Recently, one of his friends was a victim of a ransomware attack which ended up costing incident cost him tremendously. Now James thinks he should be more careful but does not know exactly what to do.

6.4 PERSONA 4: MARY

Mary is in her first semester as a freshman at the university starting a bachelor programme in computer science. Moving from high school to the university environment was an overwhelming experience at first, but now she has become accustomed to her new environment. One challenging thing for her has been the amount of time that she is on her computer to study, manage her accounts, pay her school tuition, and, of course, do her coursework. She still struggles to get all of those under control. She knows she should pay attention to protect her data and computer access, yet she does not know exactly what she needs to do.

In order to significantly, effectively, and structurally improve awareness among college students and faculty, a comprehensive course should be offered and required to all students upon their admission. By preparing and educating students to protect themselves in cyberspace, not only would we help them improve their personal security and prepare them for their professional lives, but we would also improve cybersecurity at educational institutions.

What should aware students do?

Considering all areas discussed in the previous chapters, one might ask what an aware cybersecurity student should do to protect herself/himself in cyberspace. In this section, we present a brief list of significant actions that students should take to reasonably defend themselves in the digital world.

6.5 ACCESS CONTROL

Protecting access to all devices (laptops, tablet, phone, and so on) by using a hard-to-guess password or personal identification number (PIN) is essential. In addition to a hard-to-guess access key, students could configure a shorter amount of time of inactivity before the device is locked out.

6.6 USERNAMES AND PASSWORDS

In the authentication phase for nearly all Internet accounts, creating usernames (or user identifications – IDs) is still the primary way of establishing authentication. Students, like any other users, should create hard-to-guess usernames or passwords. Hard-to-guess passwords should be used not only for bank accounts but also for all sites, including social networking sites where students have important profile information.

Users are also advised to create hard-to-guess usernames. Commonly used usernames like a first name initial, last name, or email address are too easy to guess. It is now common that the system allows a user to create a customized username that would be hard-to-guess.

Another important measure to take is to frequently change usernames and passwords. It is common knowledge that many major enterprise accounts have already been hacked and that users' data has been traded or made available on the dark web. For example, if the usernames and passwords of users have been compromised in a major case like the Yahoo hacking [99], changing the password and username can to some degree protect users from future attacks.

6.7 TWO-FACTOR AUTHENTICATION

Thus, users are required to use 2FA for some accounts, such as bank accounts. Despite the inconvenience of 2FA (i.e., the requirement to receive a text on a mobile device), it should be used whenever available in those accounts that include users' private information. Although 2FA is not a flawless security measure as it can still be compromised, it is currently an excellent security measure.

6.8 IDENTITY PROTECTION

Protecting the identity of a student from identity thefts is extremely important. When students are looking for jobs and opportunities, it is common that unknown recruiters, fake job postings, or fake job offers might ask for private information such as a social security number, student ID number, or access code to social networking. Students should, but might not, know that recruiters and unknown sources should not be asking for this type of data and that sharing this data might come at a major cost. Student ID numbers and all other very private information should only be communicated with authorized and trusted entities through password-protected portable document format (PDF) documents or encrypted channels of communication. Before sharing such vital information with anyone, students should ask themselves if that entity is authorized to have requested information and what the purpose of that information might be. For example, a recruiter does not need to have a social security number in order to receive a student's job application.

6.9 TRUST

Trust is one of the fundamental rules in cyber protection. Students should learn to analyse each situation and make a judgement to see if a source can be trusted. There are countless security questions that can and should arise in the minds of students throughout their online activity. For example, can an e-commerce site be trusted? Is the browser certificate valid? Does the site use hyper text transfer protocol secure (HTTPS)? Is an email they received legitimate? Has the email been sent by a reliable and known user that has been confirmed by checking the sender's email address?

Generally speaking, users should think of the consequences of all of their actions on the Internet.

6.10 PHISHING

Spear phishing (the fraudulent emails from a known or trusted sender to induce targeted individuals to reveal confidential information) is still an important source of attacks. Accounting to Symantec [95] spear phishing is the number one infection vector employed by 71% of organized groups in 2017. Phishing emails are becoming very sophisticated and sometimes hard for users to distinguish from standard emails or text messages. In order to maintain a safe approach and prevent phishing attacks, it is recommended for users not to click any links from an email or text message that asks them to log in in order to access an account or view a site. Students should always learn to log in from their browser to a specific site.

6.11 PRIVACY

User privacy on the Internet is probably the paramount issue above everything else. There are several types of privacy as summarized below:

- Users' private information on the Internet captured without users' consent such as cookies, locations, browsing logs, or search queries.
- Users agree to give private information to use a service provided by an application such as maps.
- Users' private information is captured, but users are not aware of the privacy policy of the application due to the hardship of reading the privacy policy or ambiguous privacy policy. For example, according to the Pew Research Center [96], most Facebook users are unaware that the social media giant is tracking their traits and interests and listing them for advertisers on its website.
- Users set the privacy settings to share information with a specific group of people on their networks.
- Users permit an app to access private particular data on their devices, for example, their location, contacts, or camera.
- Users self-disclose their private information. There are many ways in which people can divulge their personal information such as their public postings and what they share with other people or groups. This can include information

such as their political opinions, what they like or dislike, what actions they plan to take, which events they might attend, or pictures and videos containing their information.

Considering all the above cases, it is essential that students

- Read privacy policies and then permit an app to access particular data or use services.
- Set their privacy settings when using a browser or on social media. Limited sharing is highly recommended.
- Should assume that the data might not be entirely private and then they can decide if they truly want to share that information, whenever posting any information on social media.
- Never permit an app to access their private information on a device like adding applications or obtaining their data.
- Finally should assume that hackers or other actors might access any information on the Internet. Thus, before they take part in any activities like texting or email, they should know that this information might be compromised one way or another.

6.12 HOME NETWORKING

The college students, to get access to the Internet when living in the apartments or even in some housing facilities purchase, subscribe to an Internet service provider. Consequently, they use modems and routers to set their network and wireless Wi-Fi. Quite often, the service is shared among several roommates. One of the common issues is not protecting access to their router's administrator username and password or forgetting to change the manufacturer's default username, password, and IP address of the device. This leaves their routers very vulnerable to potential attackers. Consequently, it is recommended not only to password protect the wireless network but also to reset the administrator username and password and preferably change the default IP address.

6.13 MOBILE DEVICES

College students are very frequent mobile phone and tablet users due to their accessibility and mobility. Students download a variety of apps to explore an app. In addition to setting an access key, it is essential that when downloading apps, students monitor the access permissions for each application. Some access permission lists might not be clear and thus might give access to text messages, contact lists, or other information. Also, it is preferable to close applications that are not being used.

6.14 LOCATION SERVICES

Location services allows devices and third-party apps and websites to gather and use information based on the current location of the mobile devices. However, these companies might get much more than one might think from users' location data and also might sell the data to other third parties. According to an investigation by the *New York Times* [97], "At least 75 companies receive anonymous, precise location data from apps whose users enable location services to get local news and weather or other information. Several of those businesses claim to track up to 200 million mobile devices in the United States—about half those in use last year." Consequently, it is essential for all users (including students) to monitor or manage the sharing of their location settings.

It is difficult to know whether location data companies are tracking users' phones. One might assume that any app that collects location data may share the user information with other companies, as long as it mentions that somewhere in its privacy policy. To stop location tracking, users should check the internal settings in their devices and set that they do not want their location used for targeted advertisement or to disable location services altogether. Users should not permit any app to access their location if it is not needed.

6.15 PROTECTION AGAINST RANSOMWARE

Students, like other users, might be the victims of ransomware. In such an attack, once a student's file is encrypted and not accessible, they have only two options: (1) pay the ransom at the risk of not even getting their files back or (2) ignore the assets that might cost them a lot especially if they lose their course works, project, and papers. The only real protection is to back up files, pictures, videos, or other digital assets on a secure cloud backup folder or an external hard drive. If the external hard drive is used for a backup, then the hard drive must be disconnected from the computer and should not be online; otherwise, the external drive is vulnerable to ransomware attacks and could also be encrypted. If the user is a victim of ransomware and has backed up their files, then they can recover the files from the backup storage.

6.16 DELETING DATA FROM SOCIAL NETWORKING SITES

Deleting data from social networking sites or deleting a social networking account is not eliminating the previous data files that social networking shared with other vendors or even from their data files. Students are encouraged to consider this before sharing any information.

6.17 SURVEILLANCE

It is now common knowledge for many people that surveillance cameras are everywhere, on university campuses, on the street, and in shopping malls to name just a few. As a result, the images and activities of all people can be monitored and retrieved. The term "surveillance state" has become reality since the state collects information on everyone, without regard to innocence or guilt, and pretends it is not surveillance. All citizens and students should have awareness about this situation in each society. With expansion and low cost of surveillance cameras, people are also overwhelmingly using them in their personal environments and households, thus extending the surveillance state into their private lives.

Regardless of what we might think about surveillance cameras being good or bad, we must recognize that these cameras are not necessarily secure and might be hacked and transformed into spying cameras that reveal all aspects of a person's life. There are many cases of this that have been reported by the media. For example, in one case, hackers took over two-thirds of Washington, DC, police's surveillance cameras days before the 2017 presidential inauguration [98]. In another case, a zero-day vulnerability present in security camera and surveillance equipment software impacted hundreds of thousands of devices worldwide since hackers could remotely execute code in the software and replace surveillance camera feeds [100].

Thus, all students using their cameras for different purposes should be aware that without proper security, their lives could become a reality TV show for potential hackers.

6.18 TEXTING AND EMAIL

Email and text messaging are now essential methods of communicating with others. Email messages or text messages are not encrypted. Encryption protects potentially sensitive information from being read by anyone other than intended recipients. Email encryption often includes authentication. Most messaging services are now end-to-end encrypted, including Skype, WhatsApp, Facebook Messenger, and iMessage for Apple users. In order to ensure good security in email or text messaging, it is recommended to encrypt all messages that are sent or received or to use services or settings that offer end-to-end encryption and not to submit any personal information on unencrypted texts or emails.

6.19 UPDATING APPS OR OPERATING SYSTEMS AND DELETING UNUSED APPS

One of the issues that many people ignore is not updating apps on their mobile devices and computer operating systems. Users who do not update the software on their devices or applications take very high-security risks. Many applications release security updates following the discoveries of security vulnerabilities. Thus, one of the significant actions of users who are most aware of their security is to update all software, apps, or operating systems of their computers, devices, routers, cameras, or internet of things (IoT) requirements.

It is also highly recommended to delete all apps that are unused or not maintained. Apps that are in the official app stores (Apple® AppStore and Google Play™) usually follow strict development criteria. They vet the applications for things like malware. However, third-party app stores may not apply the same level of scrutiny towards those apps they allow to be listed in their stores. Thus, those apps can infect users' mobile device with malicious codes like ransomware and adware. As a result, it is advised to only download apps from official app stores.

The above list is some significant areas that an aware student should consider to protect themselves. Having said that, cybersecurity is a constantly evolving field with frequent new discoveries and issues. All users should keep a close eye on new vulnerability cases and apply the proper security actions in order to prevent any crimes or damage.

References

1. Hern A. (2018): "Marriott hotels: Data of 500m guests may have been expose," *The Guardian*, November 30, 2018. www.theguardian.com/world/2018/nov/30/marriott-hotels-data-of-500m-guests may-have-been-exposed.

2. Magee T. (2018): "British Airways' summer of failure," *Computerworlduk.com*, September 7, 2018. www.computerworlduk.com/it-business/british-airways-summer-of-failure-3683386/.

3. Lee M. (2018): "Hackers swipe data on 2 million T-mobile subscribers," *Forbes*, August 18, 2018. www.forbes.com/sites/leemathews/2018/08/24/t-mobile-hackers-swipe-data-on-2-million-subscribers/#6739f7dc7a52.

4. Granville K. (2018): "Facebook and Cambridge Analytica: What you need to know as fallout widens," *New York Times*, March 19, 2018. www.nytimes.com/2018/03/19/technology/facebook-cambridge-analytica-explained.html.

5. White Gillian N. (2017): "A cybersecurity breach at equifax left pretty much everyone's financial data vulnerable," *The Atlantic*, September 7, 2017. www.theatlantic.com/business/archive/2017/09/equifax-cybersecurity-breach/539178/.

6. Cuthbertson A. (2017): "Ransomware attacks rise 250 percent in 2017, Hitting U.S. Hardest," *Newsweek*, September 28, 2017. www.newsweek.com/ransomware-attacks-rise-250-2017-us-wannacry-614034.

7. Reyns W.B. (2010): "Stalking in the twilight zone: Extent of cyberstalking victimization," *Journal Deviant Behavior*, 33(1). www.tandfonline.com/doi/abs/10.1080/01639625.2010.538364.

8. Farzan A. (June 9, 2015): "College students are not as worried as they should be about the threat of identity theft," *Business Insider*. www.businessinsider.com/students-identity-theft-2015-6.

9. Dakss B. (2007): "College students prime target for ID theft," *CBS News*, August 21, 2007. www.cbsnews.com/news/college-students-prime-target-for-id-theft/.

10. Slusky L. and Partow-Navid P. (July 7, 2014): "Students information security practices and awareness," *Journal of Information Privacy and Security*, 8(4), 3–26. www.tandfonline.com/doi/abs/10.1080/15536548.2012.10845664.

11. Al-Janabi S. and Al-Shourbaji I. (2016): "A study of cyber security awareness in educational environment in the middle east," *Journal of Information & Knowledge Management*, 15, 1650007, 30 pages. doi:10.1142/S0219649216500076.

12. Hossain A. and Zhang W. (2016): "Privacy and security concern of online social networks from user perspective," *International Conference on Information Systems Security and Privacy (ICISSP), 2015*, Angers, pp. 246–253.

13. Senthilkumar K. and Easwaramoorthy S. (2017): "A survey on cyber security awareness among college students in Tamil Nadu," *IOP Conference Series: Materials Science and Engineering*, Volume 263, Computation and Information Technology, Tamil Nadu, pp. 1–10.

14. Grainne H. et al. (October 2017): "Factors for social networking site scam victimization among Malaysian students," *Cyberpsychology, Behavior, and Social Networking*. doi:10.1089/cyber.2016.0714.

15. Taddicken M. (January 1, 2014): "The 'Privacy Paradox' in the social web: The impact of privacy concerns, individual characteristics, and the perceived social relevance on different forms of self-disclosure," *Journal of Computer-Mediated Communication*, 19(2), 248–273. doi:10.1111/jcc4.12052.

16. Chen J., Ping W., Xu Y., and Tan B. (August 2015): "Information privacy concern about peer disclosure in online social networks," *IEEE Transactions on Engineering Management*, 62(3), 311–324.

17. Liang K., Liu J., Lu R., and Wong D. (March–April 2015): "Privacy concerns for photo sharing in online social networks," *IEEE Internet Computing*, 19(2), 58–63.

18. Li Y., Zheng N., Wang H., Sun K., and Fang H. (2017): "A measurement study on Amazon wishlist and its privacy exposure," *IEEE International Conference on Communications (ICC)*, 2017, Paris, pp. 1–7.

19. Will M., Garae J., Tan Y., Scoon C., and Ko R. (2017): "Returning control of data to users with a personal information crunch—A J. Position Paper," *IEEE TrustCom-16, The 15th IEEE International Conference on Trust, Security and Privacy in Computing and Communications.* https://markwill.me/publications/#bigcrunch.

20. Harikant N. and Suma V. (June 15–16, 2017): "Risk analysis in Facebook based on user anomalous behaviors," *International Conference on Intelligent Computing and Control Systems (ICICCS)*, Madurai, pp. 967–971. https://ieeexplore.ieee.org/document/8250609.

21. Nolte C. (2018): "California's high tech, diverse population are its new story," *San Francisco Chronically*, July 21, 2018. www.sfchronicle.com/bayarea/nativeson/article/California-s-high-tech-diverse-population-are-13093487.php.

22. San Jose State University: "Institutional effectiveness and analytics," *San Jose State University*, Accessed November 2018. www.iea.sjsu.edu/Students/QuickFacts/default.cfm?version=graphic.

23. Kruger H. and Kearney W. (2006): "A prototype for assessing information security awareness," *Computers & Security*, 25(4), 289–296.

24. Schwartz J. (2017): "Report: 7 in 10 employees struggle with cyber awareness," *mediapro.com*. www.mediapro.com/blog/2017-state-privacy-security-awareness-report/.

25. Moallem A. (2017): "Do you really trust "Privacy Policy" or "Terms of Use" agreements without reading them?" In D. Nicholson (Ed.), *Advances in Human Factors in Cybersecurity*, Springer, Cham, pp. 290–295.

26. Haggerty J. et al. (2015): "Hobson's choice: Security and privacy permissions in android and iOS devices," In T. Tryfonas and I. Askoxylakis (Eds.), *Human Aspects of Information Security, Privacy, and Trust*, Springer International Publishing, Switzerland, pp. 506–516.

27. Govani T. and Pashley H. (2009): "Student awareness of the privacy implications when using Facebook". http://lorrie.cranor.org/courses/fa05/tubzhlp.pdf.

28. Boehmer J. et al. (2014): "Determinants of online safety behaviour: Towards an intervention strategy for college students," *Behaviour & Information Technology*, 34(10), 1022–1035. https://scholars.opb.msu.edu/en/publications/determinants-of-online-safety-behaviour-towards-an-intervention-s-3.

29. Google. (2016): "How people use their devices. What marketers need to know," *Google*. https://storage.googleapis.com/think/docs/twg-how-people-use-their-devices-2016.pdf.

30. Howard J. (2016): "Americans devote more than 10 hours a day to screen time, and growing," *CNN*, July 29, 2016. www.cnn.com/2016/06/30/health/americans-screen-time-nielsen/index.html.

31. Adams A. and Sasse M. (1999): "Users are not the enemy," *Communications of the ACM*, 42(12), 41–46, reprinted (2005) In Cranor and Garfinkel (Eds.), *Security and Usability*, O'Reilly, pp. 639–649 [chapter 32].

32. Gunson N., Marshall D., Morton H., and Jack M. (2011): "User perceptions of security and usability of single- factor and two-factor authentication in automated telephone banking," *Computers & Security*, 30(4), 208–220.

33. Federal Trade Commission. (2017): "Consumer sentinel network data book," For January–December 2016, www.ftc.gov/system/files/documents/reports/consumer-sentinel-network-data-book-january-december-2016/csn_cy-2016_data_book.pdf.

34. Tatham M. (2018): "Identity theft statistics," *Experian*, March 15, 2018. www.experian.com/blogs/ask-experian/identity-theft-statistics/.

35. Shahani A. (2015): "Theft of social security numbers is broader than you might think," *NPR, All things Considered*, June 15, 2015 www.npr.org/sections/alltechconsidered/2015/06/15/414618292/theft-of-social-security-numbers-is-broader-than-you-might-think.

36. Statista. (2017): "Annual number of ransomware attacks worldwide from 2014 to 2017 (in millions)," *Statista.com.* www.statista.com/statistics/494947/ransomware-attacks-per-year-worldwide/.

37. European Union. (2018): "Data protection". http://ec.europa.eu, 2018. http://ec.europa.eu/ipg/basics/legal/data_protection/index_en.htm.

38. Miyazaki A.D. (2008): "Online privacy and the disclosure of cookie use: Effects on consumer trust and anticipated patronage," *Journal of Public Policy & Marketing*, 27(1), 19–33.

39. Englehardt S., Reisman D., Eubank C., Zimmerman P., Mayer J., Narayanan A., and Felten E.W. (May 2015): "Cookies that give you away: The surveillance implications of web tracking," *Proceedings of the 24th International Conference on World Wide Web*, ACM, Florence, May 18–22, pp. 289–299.

40. Tirtea R., Castelluccia C., and Ikonomou D. (2011): "Bittersweet cookies: Some security and privacy considerations," *European Union Agency for Network and Information Security-ENISA*.

41. Rachels J. and Rachels S. (2003): "The idea of the social contract," In S. Rachels (Ed.), *The Elements of Moral Philosophy*, McGraw-Hill, New York, p. 145.

42. Gerdes H. and Mallinckrodt B. (1994): "Emotional, social, and academic adjustment of college students: A longitudinal study of retention," *Journal of Counseling and Development*, 72(3), 281–288.

43. Lapsley D.K., Rice K.G., and Shadid G.E. (1989): "Psychological separation and adjustment to college," *Journal of Counseling Psychology*, 36(3), 286–294. doi:10.1037/0022-0167.36.3.286.

44. Maksl A. and Young R. (2013): "Affording to exchange: Social capital and online information sharing," *Cyberpsychology, Behavior, and Social Networking*, 16(8), 588–592.

45. Sheldon P. (2009): "I'll poke you. You'll poke me!" Self-disclosure, social attraction, predictability and trust as important predictors of Facebook relationships," *Cyberpsychology: Journal of Psychosocial Research on Cyberspace*, 3(2).

46. Walker L.S. and Wright P.H. (1976): "Self-disclosure in friendship," *Perceptual and Motor Skills*, 42(3), 735–742.

47. Derlega V., Metts S., Petronio S., and Margulis S.T. (1993): "*Self-Disclosure*," SAGE Publications, Thousand Oaks, CA.

48. Mazer J.P., Murphy R.E., and Simonds C.J. (2007): "I'll see you on "Facebook": The effects of computer- mediated teacher self-disclosure on student motivation, affective learning, and classroom climate," *Communication Education*, 56(1), 1–17.

49. Altman I. and Taylor D.A. (1973): "*Social Penetration: The Development of Interpersonal Relationships*," Holt, Rinehart, & Winston, New York, p. 459.

50. Limperos A.M., Tamul D.J., Woolley J.K., Spinda J.S., and Shyam Sundar S. (2014): "'It's not who you know, but who you add:' An investigation into the differential impact of friend adding and self- disclosure on interpersonal perceptions on Facebook," *Computers in Human Behavior*, 35, 496–505.

51. Turban E., Bolloju N., and Liang T.P. (2011): "Enterprise social networking: Opportunities, adoption, and risk mitigation," *Journal of Organizational Computing and Electronic Commerce*, 21, 202–220.

52. Statista. (2018): "Facebook: Number of monthly active users worldwide 2008-2018," *www. Statista.come*. www.statista.com/statistics/264810/number-of-monthly-active-facebook-users-worldwide/.

53. Noyes D. (2017): "The top 20 valuable Facebook statistics–Updated March 2017," *Zephoria Digital Marketing*, Retrieved from https://zephoria.com/top-15-valuable-facebook-statistics/.

54. Palmer A. (2018): "How much is YOUR data worth? In wake of Facebook's massive privacy scandal, experts say login details sell for just $5.20 on the dark web," *Daily Mail*, March 22, 2018. www.dailymail.co.uk/sciencetech/article-5533871/How-Facebook-data-worth-Hackers-sell-dollars.html.

55. Lamagna M. (2018): "Spooked by the Facebook privacy violations? This is how much your personal data is worth on the dark web," Retrieved from www.marketwatch.com/story/spooked-by-the-facebook-privacy-violations-this-is-how-much-your-personal-data-is-worth-on-the-dark-web-2018-03-20.

56. Wrenn E. (2012): "Half of Facebook users accept 'friend requests' from strangers, while 13m U.S. users have never opened their privacy settings," *Daily Mail*, March 4, 2012. Retrieved from www.dailymail.co.uk/sciencetech/article-2139424/Half-Facebook-users-accept-friend-requests-strangers-13m-U-S-users-NEVER-opened-privacy-settings.html.

57. Gross R. and Acquisti A. (2005): "Information revelation and privacy in online social networks," *Proceedings of the 2005 ACM workshop on Privacy in the electronic society*, ACM, Alexandria, VA, November 7–10, pp. 71–80.

58. Gao H., Hu J., Huang T., Wang J., and Chen Y. (July–August 2011): "Security issues in online social networks," *IEEE Internet Computing*, 15(4), 56–63.

59. Wang P., Angarita R., and Renna I. (2018): "Is this the era of Misinformation yet? Combining social bots and fake news to deceive the masses," *2018 Web Conference Companion*, Lyon, March 2018. https://hal.archives-ouvertes.fr/hal-01722413/file/deceptiveSocialBots.pdf.

60. Statista: "Facebook - Statistics & facts." Accessed on August 30, 2018. www.statista.com/topics/751/facebook/.

61. Gramlich J. (2018): "5 facts about Americans and Facebook," *Pew Research*, April 2018 www.pewresearch.org/fact-tank/2018/04/10/5-facts-about-americans-and-facebook/.

62. Nadkarni A. and Hofmann S.G. (February 2012): "Why do people use Facebook?" *Personality and Individual Differences*, 52(3), 243–249. www.sciencedirect.com/science/article/pii/S0191886911005149.

63. Smith A. and Anderson M. (2018): "Social media use in 2018," *Pew Research Center*, March 1, 2018. www.pewinternet.org/2018/03/01/social-media-use-in-2018/.

64. Granvillem K. (2018): "Facebook and Cambridge Analytica: What you need to know as fallout widens," *New York Times*, March 19, 2018. www.nytimes.com/2018/03/19/technology/facebook-cambridge-analytica-explained.html.

65. Liu Y., Krishnamurthy B., Gummadi K.P., and Mislove A. (2011): "Analysing Facebook privacy settings: Expectation vs reality," *IMC'11*, November 2–4, 2011, Berlin. https://conferences.sigcomm.org/imc/2011/docs/p61.pdf.

66. Mislove A., Gummadi K.P., Viswanath B., and Cha M. (2009): "On the evolution of user interaction on Facebook," *WOSN'09*, August 17, 2009, Barcelona. https://conferences.sigcomm.org/sigcomm/2009/workshops/wosn/papers/p37.pdf.

67. Tufechi Z. (2012): "Facebook, youth and privacy in networked publics," *Proceedings of the Sixth International AAAI Conference on Weblogs and Social Media*, Dublin, June 4–7, 2012. The AAAI Press, Palo Alto, CA.

68. Debatin B., Lovejoy J.P., and Horn A.-K. (2009): "Facebook and online privacy: Attitudes, behaviours and unintended consequences," *Journal of Computer Mediated Communication*, 1(15), 83–108.

69. Fan J. and Zhang P. (2011): "Study on e-government information misuse based on General Deterrence Theory," *ICSSSM'11, Tianjin, IEEE*, pp. 1–6.

70. Ng B.-Y., Kankanhalli A., and Xu Y.C. (2009): "Studying users' computer security behavior: A health belief perspective," *Decision Support Systems*, 46(4), 815–825.

71. Egelman S. and Peer E. (2015): "Scaling the security wall: Developing a Security Behavior Intentions Scale (SeBIS)," *ACM Human Factors in Computing Systems*, Seoul, pp. 2873–2882.

72. Egelman S., Harbach M., and Peer E. (2016): "Behavior ever follows intention? A validation of the Security Behavior Intentions Scale (SeBIS)," *Proceedings of the 2016 CHI Conference on Human Factors in Computing Systems*, San Jose, CA, May 7–12, pp. 5257–5261.

73. Jang-Jaccard J. and Nepal S. (2014): "A survey of emerging threats in cybersecurity," *Journal of Computer & System Sciences*, 80(5), 973–993.

74. Pahnila S., Siponen M., and Mahmood A. (2007): "Employees' behavior towards IS security policy compliance," *Hawaii International Conference on System Sciences*, Waikoloa, HI.

75. Abawajy J. (2012): "User preference of cyber security awareness delivery methods," *Behaviour & Information Technology*, 2(4), 1–12.

76. Stanton J.M., Stam K.R., Mastrangelo P., and Jolton J. (2005): "Analysis of end user security behaviors," *Computers & Security*, 24(2), 124–133.

77. Furnell S. and Clarke N. (2012): "Power to the people? The evolving recognition of human aspects of security," *Computers & Security*, 31(1), 983–988.

78. Hight S.D. (2005): "The importance of a security, education, training and awareness program (November 2005)". https://pdfs.semanticscholar.org/7f8f/ec0fb611a86e6ea33422a6268e-1ab69a2d81.pdf?_ga=2.166734973.2122401904.1540690587-1839534596.1540242606.

79. Talib S., Clarke N.L., and Furnell S.M. (2010): "An analysis of information security awareness within home and work environments," *International Conference on Availability, Reliability, and Security*, Krakowska Akademia, Poland.

80. Ponemon Institute. (2017): "Cost of CyberCrime Study, Insights on the Security investment that make a difference," *Ponemon Institute*, www.accenture.com/t20170926T072837Z__w__/us-en/_acnmedia/PDF-61/Accenture-2017-CostCyberCrimeStudy.pdf.

81. Goode J. (2018): "Comparing training methodologies on employee's cybersecurity countermeasures awareness and skills in traditional vs. socio-technical programs," CEC Theses and Dissertations, Nova Southeastern University College of Engineering and Computing.

82. Valentine J.A. (2006): "Enhancing the employee security awareness model," *Computer, Fraud & Security*, 6, 17–19.

83. Leach J. (2003): "Improving user security behavior," *Computers & Security*, 22(8), 685–692.

84. Cone B.D. et al. (2007): "A video game for cyber security training and awareness," *Computers & Security*, 26(1), 63–72.

85. Albrechtsen E. and Hovden J. (2010): "Improving information security awareness and behaviour through dialogue, participation and collective reflection. An intervention study," *Computer and Security*, 29(4), 432–445.

86. Egelman S. (2015): "Scaling the security wall: Developing a security behavior intentions scale (SeBIS)," *CHI '15 Proceedings of the 33rd Annual ACM Conference on Human Factors in Computing Systems*, Seoul, April 18–23, pp. 2873–2882.

87. Egelman et al. (2016): "Behavior ever follows intention?: A validation of the security behavior intentions scale (SeBIS)," *CHI '16 Proceedings of the 2016 CHI Conference on Human Factors in Computing Systems*, San Jose, CA, May 7–12, pp. 5257–5261.

88. Parsons K. et al. (January 2017): "The Human Aspects of Information Security Questionnaire (HAIS-Q): Two further validation studies," *Computers & Security*, 66, 40–51.

89. Albrechtsen A. (June 2007): "A qualitative study of users' view on information security," *Computers & Security*, 26(4), 276–289.

90. Tara E. (2017): "Cost of user security training tops $290K per year," *infosecurity-magazine.com*, September 15, 2017. www.infosecurity-magazine.com/news/cost-of-user-security-training/.

91. Pascual A., Marchini K., and Miller S. (2018): "Identity fraud: Fraud enters a new era of complexity," *javelinstrategy.com*, February 6, 2018. www.javelinstrategy.com/coverage-area/2018-identity-fraud-fraud-enters-new-era-complexity.

92. Cook S. (2018): "Ransomware statistics and facts in Antivirus 2017-2018," *Comparitech.com*, August 25, 2018. www.comparitech.com/antivirus/ransomware-statistics/.

93. Alotaibi F. et al. (2016): "A review of using gaming technology for cyber-security awareness," *International Journal for Information Security Research (IJISR)*, 6(2), 660–666.

94. Hendrix M., Al-Sherbaz A., and Bloom V. (2016): "Game based cyber security training: Are serious games suitable for cyber security training?" *International Journal of Serious Games*, 3(1), 53–61.

95. Symantec. (2018): "Internet security threat report (ISTR)," Volume 23, Symantec, Corporation. http://images.mktgassets.symantec.com/Web/Symantec/%7B3a70beb8-c55d-4516-98ed-1d0818a42661%7D_ISTR23_Main-FINAL-APR10.pdf?aid=elq_.

96. Gramlich J. (2019): "10 facts about Americans and Facebook," *Pew Research Center, FACTANK*, February 1, 2019. www.pewresearch.org/fact-tank/2019/02/01/facts-about-americans-and-facebook/.

97. Valentino-DeVaries et al. (2018): "Your apps know where you were last night, and they're not keeping it secret," *New York Times*, December 10, 2018. www.nytimes.com/interactive/2018/12/10/business/location-data-privacy-apps.html.

98. Hermann P. (2018): "Hack of D.C. police cameras was part of ransomware scheme, prosecutors say," *Washington Post*, July 28, 2018. www.washingtonpost.com/local/public-safety/attack-on-dc-police-security-cameras-had-broad-implications/2018/07/24/7ff01d78-8440-11e8-9e80-403a221946a7_story.html?utm_term=.c123fa9ceb94.

99. Greeberg A. (2016): "Hack brief: Yahoo breach hits half a billion users," *WIRED*, September 22, 2016. www.wired.com/2016/09/hack-brief-yahoo-looks-set-confirm-big-old-data-breach/.

100. Osborne C. (2018): "Hackers hijack surveillance camera footage with 'Peekaboo' zero-day vulnerability," *Zero Day*, September 17, 2018. www.zdnet.com/article/hackers-can-tamper-with-surveillance-camera-footage-due-to-new-zero-day-vulnerability/.

Glossary

Antivirus is a software designed to detect and destroy computer viruses.

Authentication is the process of establishing the identity of a user through electronic methods.

Bitcoin (BTC) is a type of digital currency that is created and held electronically. It is not printed and is not a tangible form of currency. It is not controlled or regulated, and people and, increasingly, businesses using software that solves mathematical problems on computers all over the world produce it.

Cache sniffing is capturing, or "sniffing," the browser cache and history of a visitor and using them to deceive the user more accurately.

Clickjacking is a technique of tricking a user into clicking on something different from what the user perceives, thus potentially revealing confidential information or allowing others to take control of their computer.

Cookie sniffing is capturing, or "sniffing cookies of users" computer then used to capture, for example, a session cookie or username/password.

Crawler is a software program that systematically browses the World Wide Web to create an index of data.

Cryptocurrency is a digital currency built with cryptographic protocols that make transactions secure, difficult to fake, and not controlled by any central authority.

CryptoLocker is a malware threat that infects a computer and then searches for files to encrypt. CryptoLocker encrypts anything on hard drives and all connected media, for example, USB memory sticks or any shared network drives.

Cryptology or cryptography is the science of constructing and analysing protocols that prevent third parties or the public from reading private messages. It includes the design of various ciphers, cryptanalysis methods (attacks), key exchange, key authentication, cryptographic hashing, digital signing, and social issues.

Cyberbullying is the use of electronic information and communication devices such as emails, instant messaging, text messages, mobile phones, pagers, and defamatory websites to harass an individual or group through personal attacks or other means.

Cyber-stalking is the use of technology, particularly the Internet, to harass someone.

Data mining is the process of discovering patterns and correlations within large data.

Federal Bureau of Investigation (FBI) is a U.S. agency in charge of domestic intelligence and internal security. It is operating under the jurisdiction of the U.S. Department of Justice.

Federal Trade Commission (FTC) is a U.S. agency that provides information to help consumers identify, prevent, and avoid scams and fraud, and works to stop fraudulent, deceptive, and unfair business practices.

Global positioning system (GPS) is a global navigation satellite system that provides geo-location and time information to a GPS receiver anywhere on or near the Earth. The U.S. government owns GPS, and the U.S. Air Force operated it.

Google Play is a digital distribution service operated and developed by Google LLC. It serves as the official app store for the Android operating system.

Hyper Text Transfer Protocol Secure (HTTPS) is the secure version of HTTP, the protocol over which data is sent between the browser and the website.

Hypertext Markup Language (HTML) is the standard markup language for creating web pages and web applications.

Identity clone attack is an attack that attacker deceives the user's friend to make a healthy relationship with him by replicating the user's identity either in the same network or in another network.

Identity theft is when thieves steal someone's personal information to take over or open new accounts, file fake tax returns, rent or buy properties, and so on.

Internet cookies are messages that web servers pass to the web browser when the user visits Internet sites. The browser then stores each message in a small file, called cookie.txt. When users request another page from the server, the browser sends the cookie back to the server.

Internet of Things (IoTs) are physical devices, vehicles (also referred to as "connected devices" and "smart devices"), buildings, and other items embedded with electronics, software, sensors, actuators, and network connectivity that are able to collect and exchange data.

Internet Protocol (IP) address is a numerical label assigned to each device connected to a computer network that uses the Internet Protocol for communication.

iOS is a mobile operating system created and developed by Apple Inc.

Learning Management System is a software application for the administration, documentation, tracking, reporting, and delivery of educational courses, training programmes, or learning and development programmes.

Linux is a family of open-source software operating systems based on the Linux kernel.

Malware is software designed to disrupt, damage, or gain unauthorized access to a computer system.

OS command injection is an attack technique used for unauthorized execution of operating system commands.

Phishing is a fraudulent attempt to obtain sensitive information such as usernames, passwords, and credit card details by disguising as a trustworthy entity in electronic communication.

Python is a programming language.

Quick Response Code (QR code) is a machine-readable optical label that contains information about the item to which it is attached.

Ransomware is malicious software from cryptovirology that threatens to publish the victim's data or perpetually block access to it unless a ransom is paid.

Secure Socket Layer (SSL) is the standard security technology for establishing an encrypted link between a web server and a browser.

Social Contract Theory (SCT) is the view that person's moral and/or political obligations are dependent upon a contract or agreement.

Social engineering is "any act that influences a person to take an action that may or may not be in their best interest" (social-engineer.org 2018). "Social engineering, in the context of information security, refers to psychological manipulation of people into performing actions or divulging confidential information" (wikipedia.org 2018).

SQL injection is a code injection and is one of the most common web hacking techniques. The attacker places malicious code in SQL (Structured Query Language) statements, via web page input that might destroy your database.

Sybil attacks are attacks wherein a reputation system is subverted by forging identities in peer-to-peer networks.

TOR browser or "the onion routing" is a web browser designed for anonymous web surfing and protection against traffic analysis.

Two-factor authentication (2FA) is a type of authentication that requires the confirmation of users' claimed identities by using a combination of any two different factors: (1) something they know, (2) something they have, or (3) something they are.

Uniform resource locator (URL) is a reference to a web resource that specifies its location on a computer network and a mechanism for retrieving it.

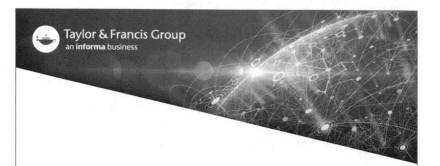